机电英语

主编 屈南可 鹿璐

高等教育出版社·北京

内容简介

本书是职业院校机电技术应用专业英语基础教材,采用"校企合作、理实一体"的项目教学模式编写,内容设计注重实践性和应用性。

根据机电技术应用专业涉及的工作任务和职业能力要求,本书主要内容包括机电一体化、机械制图、常见电子元器件及电路、单片机、PLC、工业机器人、CNC简介、CNC机床、CNC操作和编程及求职,共十个单元。通过学习使学生掌握相关专业的英语表达,为后续专业知识的深入学习奠定英语基础。

本书配有二维码资源及学习卡资源,用移动设备扫描二维码即可随时随地浏览资源,享受立体化阅读体验,学习卡资源请登录Abook网站http://abook.hep.com.cn/sve获取,详细说明见本书"郑重声明"页。

本书可以作为中等职业学校及五年一贯制学校机电技术应用专业教材,也可供电子技术应用和数控技术应用等专业使用。

图书在版编目(CIP)数据

机电英语 / 屈南可,鹿璐主编. -- 北京:高等教育出版社,2020.8

ISBN 978-7-04-053940-0

Ⅰ. ①机… Ⅱ. ①屈… ②鹿… Ⅲ. ①机电工程-英语-中等专业学校-教材 Ⅳ. ①TH

中国版本图书馆CIP数据核字(2020)第050220号

策划编辑	项 杨	责任编辑	项 杨	封面设计	张 楠	版式设计	童 丹
插图绘制	于 博	责任校对	任 纳 陈 杨	责任印制	刘思涵		

出版发行	高等教育出版社	网 址	http://www.hep.edu.cn
社 址	北京市西城区德外大街4号		http://www.hep.com.cn
邮政编码	100120	网上订购	http://www.hepmall.com.cn
印 刷	山东韵杰文化科技有限公司		http://www.hepmall.com
开 本	787 mm×1092 mm 1/16		http://www.hepmall.cn
印 张	11.5		
字 数	260千字	版 次	2020年8月第1版
购书热线	010-58581118	印 次	2020年8月第1次印刷
咨询电话	400-810-0598	定 价	28.00元

本书如有缺页、倒页、脱页等质量问题,请到所购图书销售部门联系调换
版权所有 侵权必究
物料号 53940-00

前言

本书是职业院校机电技术应用专业英语基础教材，采用"校企合作、理实一体"的项目教学模式编写。在内容上以突出基本概念，注重技能训练，强调理论联系实际，加强实践性教学环节为原则，凸显实践性、应用性和层次性。

本书主要内容包括机电一体化、机械制图、常见电子元器件及电路、单片机、PLC、工业机器人、CNC简介、CNC机床、CNC操作和编程及求职，共十个单元，使学生在学习专业英语的语法特点、文体结构的基础上，掌握一定数量的专业英语词汇，培养学生以英语为工具获得更多专业知识的技能，为后续专业学习奠定英语基础。

本书主要特色如下：

1. 在内容选取上，本书涉及的机械、电子及数控机床的英语基本知识来自于专业课实际教学内容，与专业知识相辅相成；数控材料源于真实项目的实际操作，贴近生产实际。

2. 在编写形式上，本书立足于教学需要和学生实际情况，通过各种方式力求生动化、立体化地讲授相关知识。

3. 在配套资源上，充分考虑助教助学。制作了与各个单元中相关知识点配套的教学视频及每篇课文和单词的音频，通过二维码与相关内容链接，既方便学生进行跟读学习、练习听说，也使学生更加直观地了解课程内容。另外，还制作了与每课内容配套的演示文稿、电子教案及单元测试题，方便教师参考选用。

本书参考学时为80学时，各单元内容相对独立，教师可根据教学实际情况灵活安排教学内容及教学时间。

本书配套的学习卡资源包括电子教案、演示文稿、译文及练习题参考答案等，可登录Abook网站http://abook.hep.com.cn/sve获取，详细说明见本书"郑重声明"页。

本书配套红膜学习卡，覆上红膜之后，专色印刷的单词可以隐去，方便检测学习效果。

本书由屈南可、鹿璐任主编，闫少华任副主编，具体编写分工如下：屈南可编写单元1、3、5、7、9的课文及相关译文，鹿璐编写单元2、4、6、8、10的课文及相关译文，闫少华编写练习题。

由于编者水平所限，书中不当之处在所难免，敬请读者批评指正。读者意见反馈邮箱：zz_dzyj@pub.hep.cn。

编者
2019年6月

Contents

Unit 1	Mechatronics		1
	Lesson 1	Introduction to mechatronics	2
	Lesson 2	Mechanical components	6
	Lesson 3	Common metal materials	12
Unit 2	Mechanical Drawing		19
	Lesson 1	Views	20
	Lesson 2	Assembly drawing and detail drawing	26
	Lesson 3	Introduction to AutoCAD	31
Unit 3	Electronic Components and Circuits		39
	Lesson 1	Common electronic components	40
	Lesson 2	Electric circuits	44
	Lesson 3	Introduction to Multisim	49
Unit 4	Single-chip Microcomputer		55
	Lesson 1	Introduction to single-chip microcomputer	56
	Lesson 2	Applications of single-chip microcomputer	60
	Lesson 3	MCS-51™ instruction set summary	65
Unit 5	Programmable Logic Controller (PLC)		69
	Lesson 1	Introduction to PLC	70
	Lesson 2	PLC operations	75
	Lesson 3	Troubleshooting of PLC	80
Unit 6	Industrial Robots		85
	Lesson 1	Introduction to robots	86
	Lesson 2	Types of industrial robots	89
	Lesson 3	Typical applications of industrial robots	94
Unit 7	General View of CNC		99
	Lesson 1	Introduction to CNC	100
	Lesson 2	Applications of CNC	103
	Lesson 3	Safety requirements of CNC operations	108
Unit 8	CNC Machine Tools		113
	Lesson 1	CNC lathe and turning operations	114
	Lesson 2	Milling machines and milling operations	120
	Lesson 3	Machining centers	124

Contents

Unit 9　CNC Operation and Programming ·· 129
　　　　Lesson 1　Operation panel—system control panel ················ 130
　　　　Lesson 2　Operation panel—machine control panel ·············· 136
　　　　Lesson 3　Commonly used codes and control functions ········ 142
Unit 10　Job Application ·· 149
　　　　Lesson 1　Job advertisement ·· 150
　　　　Lesson 2　Personal resume ·· 152
　　　　Lesson 3　Job interview ·· 155
Word List ·· 159
Phrases and Expressuions ·· 167
References ·· 174

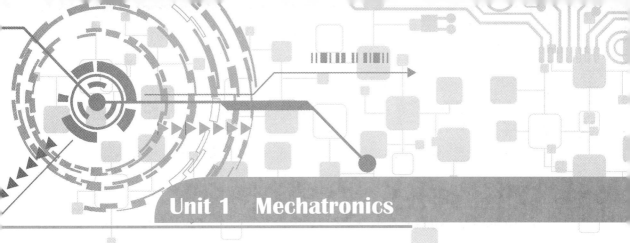

Unit 1   Mechatronics

In this unit, you will learn

- ◇ 1. Introduction to mechatronics;
- ◇ 2. Mechanical components;
- ◇ 3. Common metal materials.

Lesson 1 Introduction to mechatronics

1. What is mechatronics

The word mechatronics is composed of "mecha" from mechanics and "tronics" from electronics.

The mechatronics field is the intersection of three traditional engineering fields. They are:

(1) Mechanical engineering, where the part "mecha" is taken from;

(2) Electrical or electronics engineering, where the part "tronics" is taken from;

(3) Computer engineering.

2. Applications of mechatronics

Mechatronics has a wide range of applications as follows.

2.1 Design and modeling

Design and modeling are simplified by using mechatronic systems. There are many designing tools such as AutoCAD, through which two-dimensional (2D) drawings or three-dimensional (3D) drawings (Fig.1.1.1) can be made.

The virtual modeling of a manufacturing factory gives an idea of the time taken for a component to be manufactured and also shows virtually how the operations will be performed.

2.2 Software integration

Different kinds of software are used in manufacturing, design, testing, monitoring, and control of the manufacturing process. Examples of such software include computer aided design (CAD), computer aided testing (CAT) and computer integrated manufacturing (CIM)(Fig.1.1.2).

2.3 Intelligent control

There are many industrial processes and machines which control many variables automatically. Temperature, flow, pressure, speed, etc. are maintained constantly by controllers.

2.4 Robotics

Robot technology uses mechanical, electronic and computer systems. A robot is a multifunctional reprogrammable machine. Robots are able to perform operations, assembly, spot welding, spray

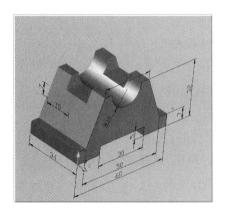

Fig.1.1.1　Three-dimensional (3D) drawing

Fig.1.1.2　CIM

painting, etc (Fig.1.1.3).

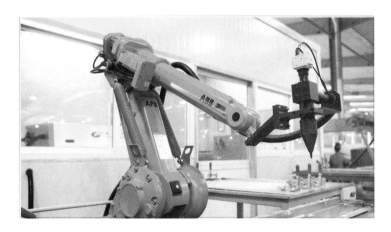

Fig.1.1.3　Robotics

New Words

1-1 单词

mechatronics [ˌmekəˈtrɔniks]	n. 机电一体化；机械电子学
mechanics [miˈkæniks]	n. 机械学
electronics [iˌlekˈtrɔniks]	n. 电子学；电子设备
intersection [ˌintəˈsekʃn]	n. 交集
engineering [ˌendʒiˈniəriŋ]	n. 工程；工程学
mechanical [miˈkænikl]	adj. 机械的；机械学的
electrical [iˈlektrikl]	adj. 电学的；与电有关的
application [ˌæpliˈkeiʃn]	n. 应用；运用

model [ˈmɔdl]	n. 模型
	vt. 制作模型
simplify [ˈsimpliˌfai]	vt. 简化
system [ˈsistəm]	n. 体系；系统
tool [tu:l]	n. 工具
dimensional [diˈmenʃənəl]	adj. 尺寸的；维的
virtual [ˈvə:tʃuəl]	adj. （计算机）虚拟的
manufacture [ˌmænjuˈfæktʃə]	vt. 制造；生产
component [kəmˈpəunənt]	n. 部件；零件
operation [ˌɔpəˈreiʃn]	n. 操作
perform [pəˈfɔ:m]	vt. & vi. 执行；履行
software [ˈsɔftweə]	n. 软件
integration [ˌintiˈgreiʃn]	n. 整合；一体化
monitor [ˈmɔnitə]	vt. 监控；测定
control [kənˈtrəul]	vt. 控制
process [ˈprəuses]	n. 过程；工序
include [inˈklu:d]	vt. 包括；包含
intelligent [inˈtelidʒənt]	adj. 智能的
variable [ˈveəriəbl]	n. 变量
automatically [ˌɔ:təˈmætikli]	adv. 自动地
flow [fləu]	n. 流量
pressure [ˈpreʃə]	n. 压力
maintain [meinˈtein]	vt. 保持；维持
constantly [ˈkɔnstəntli]	adv. 始终地；恒定地
controller [kənˈtrəulə]	n. 控制器
robotics [rəuˈbɔtiks]	n. 机器人技术
multifunctional [ˌmʌltiˈfʌŋkʃənl]	adj. 多功能的
reprogrammable [ripˈrəugræməbl]	adj. 可改编程序的；可编程的
assembly [əˈsembli]	n. 组装；装配

Phrases and Expressions

be composed of	由……组成
two-dimensional (2D) drawing	二维绘图
three-dimensional (3D) drawing	三维绘图
computer aided design (CAD)	计算机辅助设计 (CAD)
computer aided testing (CAT)	计算机辅助测试 (CAT)
computer integrated manufacturing (CIM)	计算机集成制造 (CIM)

intelligent control	智能控制
robot technology	机器人技术
spot welding	点焊
spray painting	喷漆

Exercises

I. Read and judge.

(　　) 1. The mechatronics field includes three traditional engineering fields.

(　　) 2. The mechatronics field is the intersection of the mechanical engineering, electrical engineering and computer engineering.

(　　) 3. Different kinds of software are used in manufacturing, design, testing, monitoring, and control of the manufacturing process.

(　　) 4. Temperature, flow, pressure, speed, etc. are maintained constantly by controllers.

(　　) 5. Robot technology only uses computer systems.

II. Fill in the blanks.

1. The word mechatronics is composed of "mecha" from _____ and "tronics" from _____.

2. Different kinds of software are used in _____, _____, _____, _____, and control of the manufacturing process.

3. There are many industrial processes and machines which control many _____ automatically.

4. Robots are able to perform operations, _____, _____, _____, etc.

III. List the four applications of mechatronics, and then translate them into Chinese.

1. _____   _____
2. _____   _____
3. _____   _____
4. _____   _____

IV. Give the full names of the abbreviations, and then translate them into Chinese.

Abbreviation	English	Chinese
3D		
CAD		
CAT		
CIM		

V. Translate the following sentences into Chinese.

1. The word mechatronics is composed of "mecha" from mechanics and "tronics" from electronics.

2. The mechatronics field is the intersection of three traditional engineering fields.

3. Design and modeling are simplified by using mechatronic systems.

4. There are many industrial processes and machines which control many variables automatically.

5. A robot is a multifunctional reprogrammable machine.

Lesson 2　Mechanical components

Mechanical components (Fig.1.2.1) are the basic elements to form a machine. They play important roles in mechanical engineering though they look very small compared to the whole machines.

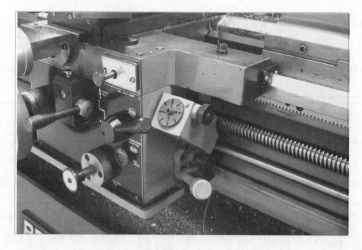

Fig.1.2.1　Mechanical components

1. Screw

A screw (Fig.1.2.2) is a type of fastener. It is used to clamp machine parts together when one of the parts has an internal thread. It can be turned with screwdrivers or wrenches. If none of the parts is threaded, a bolt must be used, which consists of a screw, a nut, and usually a washer.

2. Spring

A spring (Fig.1.2.3) is an elastic object used to store mechanical energy.

Springs are used for many purposes, such as supplying the motive power in clocks and watches, cushioning transport vehicles, measuring weights, etc.

3. Shaft

A shaft (Fig.1.2.4) is a cylindrical bar that supports and rotates with devices. It is used to transmit power from one part to another. The members such as pulleys and gears are mounted on it.

Fig.1.2.2 Screws and bolts   Fig.1.2.3 Springs   Fig.1.2.4 Shafts

4. Gear

A gear is a rotating machine part with cut teeth. Gears (Fig.1.2.5) are used to transmit motion and power from one rotating shaft to another.

5. Bearing

A bearing is a machine object that is used to bear (to support) the shaft and the objects connected to the shaft.

A ball bearing (Fig.1.2.6) is a type of rolling bearing that uses balls to maintain the separation between the bearing races. Ball bearings are used in almost every kind of machine with rotating parts. A ball bearing usually consists of four parts: an inner race, an outer race, balls and a cage.

Fig.1.2.5 Gears

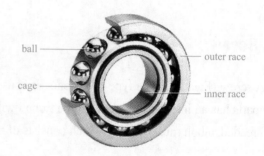

Fig.1.2.6 Ball bearing

6. Coupling

A coupling is used to connect two shafts together at their ends to transmit power. There are two general types of couplings: rigid couplings and flexible couplings.

Rigid couplings (Fig.1.2.7a) are designed to draw two shafts tightly together so that no relative motion occurs between them.

Flexible couplings (Fig.1.2.7b) permit the shafts to move "freely" in the axial direction without interfering with one another.

(a) Rigid coupling (b) Flexible couplings

Fig.1.2.7 Couplings

1-2 单词

New Words

element ['elimənt]	n. 元素；要素
screw [skru:]	n. 螺钉
fastener ['fɑ:sənə]	n. 紧固件
clamp [klæmp]	vt. & vi. 夹紧；锁住

internal [inˈtə:nl]	*adj.* 内部的
thread [θred]	*n.* 螺纹；线
screwdriver [ˈskru:ˈdraivə]	*n.* 螺丝刀
wrench [rentʃ]	*n.* 扳手
bolt [bəult]	*n.* 螺栓
nut [nʌt]	*n.* 螺母
washer [ˈwɔʃə]	*n.* 垫圈；垫片
spring [spriŋ]	*n.* 弹簧
elastic [iˈlæstik]	*adj.* 有弹力的
energy [ˈenədʒi]	*n.* 能量
cushion [ˈkuʃn]	*n.* 垫子
	vt. 缓冲
shaft [ʃɑ:ft]	*n.* 轴
cylindrical [səˈlindrikəl]	*adj.* 圆柱形的
support [səˈpɔ:t]	*vt.* 支承
rotate [rəuˈteit]	*vt. & vi.* 使旋转；转动
transmit [trænsˈmit]	*vt.* 传输；传递
pulley [ˈpuli]	*n.* 滑轮；带轮
gear [giə]	*n.* 齿轮
mount [maunt]	*vt.* 安装
bearing [ˈbeəriŋ]	*n.* 轴承
coupling [ˈkʌpliŋ]	*n.* 联轴器
rigid [ˈridʒid]	*adj.* 刚性的
flexible [ˈfleksəbl]	*adj.* 柔性的；柔韧的
axial [ˈæksiəl]	*adj.* 轴的；轴向的
interfere [ˌintəˈfiə]	*vi.* 干涉；干扰

Phrases and Expressions

consist of	由……组成；包括
ball bearing	球轴承
rigid coupling	刚性联轴器
flexible coupling	柔性联轴器

Exercises

I. Read and judge.

(   ) 1. Mechanical components are the basic elements to form a machine.

(   ) 2. A screw is a type of fastener.

(   ) 3. Springs are used to supply the motive power in clocks and watches.

(   ) 4. A ball bearing usually consists of three parts: an inner race, an outer race and balls.

(   ) 5. Rigid couplings permit the shafts to move "freely" in the axial direction.

II. Translate the following phrases.

English	Chinese
1. mechanical components	
2. motive power	
3. cut teeth	
4. flexible coupling	
5.	机械能量
6.	球轴承
7.	内螺纹
8.	刚性联轴器

III. Fill in the blanks according to the text.

1. A screw is a type of _____. It is used to clamp machine parts together when one of the parts has an _____. If none of the parts is threaded, a _____ must be used, which consists of a _____, a _____, and usually a _____.

2. A shaft is a cylindrical bar that _____ and _____ with devices.

3. Gears are used to _____ motion and power.

4. A ball bearing usually consists of four parts: _____, _____, _____ and _____.

5. There are two general types of couplings: _____ and _____.

IV. Give the English and Chinese names of the components.

1.
English name _____
Chinese name _____

2.
English name _____
Chinese name _____

3.
English name _____
Chinese name _____

4.
English name _____
Chinese name _____

5.
English name _____
Chinese name _____

6.
English name _____
Chinese name _____

7.
English name _____
Chinese name _____

V. Translate the following sentences into Chinese.

1. If none of the parts is threaded, a bolt must be used, which consists of a screw, a nut, and usually a washer.

_____

2. A spring is an elastic object used to store mechanical energy.

_____

3. Gears are used to transmit motion and power from one rotating shaft to another.

_____

4. A ball bearing usually consists of four parts: an inner race, an outer race, balls and a cage.

_____

5. A Coupling is used to connect two shafts together at their ends to transmit power.

_____

Lesson 3 Common metal materials

Engineering materials can be divided into two groups: metal materials and non-metal materials. Almost 75% of the elements are made of metal materials. All metals can be divided into ferrous metals and non-ferrous metals.

1. Ferrous metals

Ferrous metals contain iron. Iron and many of its alloys, including steels and cast irons, form the ferrous metal group. The main difference between steels and cast irons is the content of carbon. Steels have a carbon content of 0.03%~2.0%, and cast irons have a carbon content of 2.0%~4.3%.

Ferrous metals are magnetic and have poor corrosion resistance.

1.1 Low carbon steels

Low carbon steels (Fig.1.3.1) contain carbon up to 0.30%.

Low carbon steels have good formability and weldability, and they are low cost. Low carbon steels can be used as chains, nails, pipes, machine and structural parts and so on.

1.2 Medium carbon steels

Medium carbon steels (Fig.1.3.2) contain carbon from 0.30% to 0.80%.

They have a good balance of properties. They can be used as gears, shafts, screwdrivers, wrenches, cable, machine parts and so on.

Fig.1.3.1   Low carbon steels: pipes

Fig.1.3.2   Medium carbon steels: a wrench

1.3 High carbon steels

High carbon steels (Fig.1.3.3) contain carbon from 0.80% to 2.0%.

They have low formability and weldability, but high hardness and wear resistance. They can be used as springs, drills, cutting tools and so on.

1.4 Stainless steels

Stainless steels (Fig.1.3.4) are a family of corrosion resistant steels and they contain chromium at least 10.5%. The chromium in the alloy forms an oxide layer. This oxide layer gives stainless steels a corrosion resistance.

Fig.1.3.3 High carbon steels: drills

Fig.1.3.4 Stainless steels: cookware

Stainless steels have good corrosion resistance and mechanical properties. They are widely used in cookware, industrial equipment, structural buildings, etc.

2. Non-ferrous metals

Non-ferrous metals do not contain iron. They are non-magnetic and have more corrosion resistance.

2.1 Aluminum

Pure aluminum is easily alloyed with small amounts of copper and other elements. Aluminum alloys (Fig.1.3.5) are of low density, easily formed, machined or cast. They are often used as window frames, aircraft parts, automotive parts, etc.

2.2 Copper

Copper (Fig.1.3.6) is a kind of ductile metal that conducts heat and electricity well. Copper can be used as electrical wires, pipes, printed circuit boards, etc.

Fig.1.3.5 Aluminum alloys: window frame

Fig.1.3.6 Copper: electrical wire

2.3 Brass

Brass (Fig.1.3.7) is an alloy of copper and zinc. It has reasonable hardness and can be cast, formed, and machined well. It is used as parts for electrical fittings, valves, etc.

Fig.1.3.7　Brass: valve

1-3 单词

New Words

metal [ˈmetl]	n. 金属；金属元素
material [məˈtiəriəl]	n. 材料；原料
divide [diˈvaid]	vt. （使）分开；分成
non-metal [ˈnɔnˈmetl]	n. 非金属
ferrous [ˈferəs]	adj. 铁的；含铁的
non-ferrous [ˈnʌnˈferəs]	adj. 不含铁的；有色金属的
iron [ˈaiən]	n. 铁器；铁制品
	adj. 铁制的
alloy [ˈælɔi]	n. 合金
	vt. 铸成合金
steel [sti:l]	n. 钢；钢铁
	adj. 钢制的
cast [kɑ:st]	vt. 铸造
	n. 铸型
carbon [ˈkɑ:bən]	n. 碳
	adj. 碳的
magnetic [mægˈnetik]	adj. 有磁性的；磁的
corrosion [kəˈrəuʒn]	n. 腐蚀；锈蚀
resistance [riˈzistəns]	n. 阻力；电阻
formability [fɔ:məˈbiliti]	n. 可模锻性；可成形性
weldability [weldəˈbiliti]	n. 焊接性；可焊性
chain [tʃein]	n. 链条；链子
nail [neil]	n. 钉子
balance [ˈbæləns]	n. 平衡
property [ˈprɔpəti]	n. 特性；属性
cable [ˈkeibl]	n. 电缆

drill [dril]	*n.* 钻头
	vt.& vi. 钻(孔)
stainless [ˈsteinlis]	*adj.* 不锈的
chromium [ˈkrəumiəm]	*n.* 铬
cookware [ˈkukweə]	*n.* 烹饪用具;炊具
aluminum [əˈlju:minəm]	*n.* 铝
copper [ˈkɔpə]	*n.* 铜
density [ˈdensəti]	*n.* 密度
automotive [ˌɔ:təˈməutiv]	*adj.* 自动的;汽车的
ductile [ˈdʌktail]	*adj.* 可延展的;有韧性的
circuit [ˈsə:kit]	*n.* 电路;线路
brass [brɑ:s]	*n.* 黄铜
zinc [ziŋk]	*n.* 锌
valve [vælv]	*n.* 阀门

Phrases and Expressions

corrosion resistance	耐(腐)蚀性;耐蚀力
oxide layer	氧化层
printed circuit board	印制电路板

Exercises

I. Read and judge.

(　　) 1. Engineering materials can be divided into two groups: metal materials and non-metal materials.

(　　) 2. All metals can be divided into ferrous metals and non-ferrous metals.

(　　) 3. Ferrous metals do not contain iron.

(　　) 4. Low carbon steels contain carbon from 0.80% to 2.0%.

(　　) 5. Non-ferrous metals are non-magnetic and have less corrosion resistance.

II. Translate the following phrases.

English	Chinese
1. corrosion resistance	
2. formability and weldability	
3. low carbon steels	

	continued
English	**Chinese**
4. printed circuit board	
5.	氧化层
6.	高硬度
7.	不锈钢
8.	含碳量

III. Fill in the blanks according to the text.

1. Engineering materials can be divided into two groups: _____ materials and _____ materials. All metals can be divided into _____ metals and _____ metals.

2. Ferrous metals contain _____. The main difference between steels and cast irons is _____. Ferrous metals are _____ and have _____.

3. _____ contain carbon from 0.80% to 2.0%.

4. _____ are a family of corrosion resistant steels and they contain _____ at least 10.5%.

5. _____ is a kind of ductile metal that conducts heat and electricity well.

6. _____ is an alloy of copper and zinc.

IV. Fill in the blanks according to the text.

1.　　　　　　　　2.　　　　　　　　3.　　　　　　　　4.

material: _____　_____　_____　_____
name: _____　_____　_____　_____

5.　　　　　　　　6.　　　　　　　　7.

material: _____　_____　_____
name: _____　_____　_____

V. Complete the form according to the text.

Name	Content of carbon	Property	Products
low carbon steels			
medium carbon steels			
high carbon steels			

VI. Translate the following sentences into Chinese.

1. The main difference between steels and cast irons is the content of carbon.

2. Stainless steels are a family of corrosion resistant steels and they contain chromium at least 10.5%.

3. Pure aluminum is easily alloyed with small amounts of copper and other elements.

4. Copper is a kind of ductile metal that conducts heat and electricity well.

5. Brass is an alloy of copper and zinc.

Unit 2　Mechanical Drawing

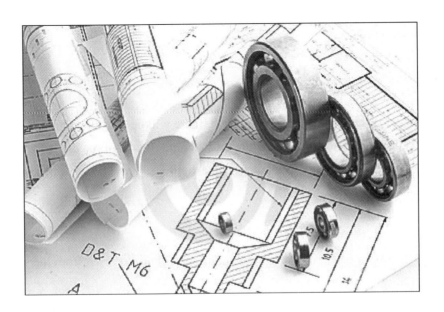

In this unit, you will learn

◇ 1. Views;
◇ 2. Assembly drawing and detail drawing;
◇ 3. Introduction to AutoCAD.

Lesson 1 Views

1. Mechanical drawing

Mechanical drawing refers to using drawings to describe how a product is made or assembled. These types of drawings represent the physical shapes and sizes of the items they describe. The basic content is to draw to scale to represent the actual objects.

In the area of engineering drawing, many types of views are used to express the design ideas in mechanical drawing.

2. Views

When an object is in the sun or under the light, there will be a shadow on the floor or on the wall. This is called projection. The drawings of machine parts by projection are called views. Views just show the visible parts of a component; the invisible parts are drawn out only when necessary.

2.1 Projection view

Projection view is an orthographic projection of a machine part seen from the sides of front, top, right, etc.

- ***Basic views (Fig.2.1.1)***

According to the regulations of the national standard mechanical drawing, the six views resulted from the projections of the machine parts on six basic planes of a cube are called basic views.

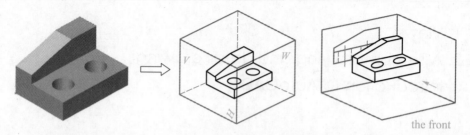

Fig.2.1.1 Basic views

Basic views and the directions of projections (Fig.2.1.2) are specified as follows:

Main view—projected from the front to the back.

Top view—projected from the top to the bottom; placed below the main view.

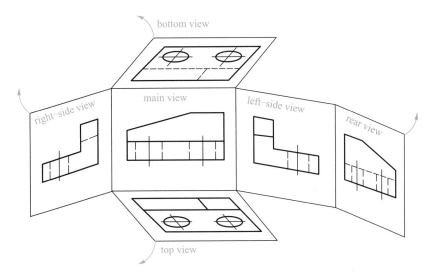

Fig.2.1.2 Six basic projection views

Left-side view—projected from the left to the right; placed on the right to the main view.
Bottom view—projected from the bottom to the top; placed above the main view.
Right-side view—projected from the right to the left; placed on the left to the main view.
Rear view—projected from the back to the front; placed on the right to the left-side view.

- ***Three-view drawing***

The main view, the top view and the left-side view are often used in drawing machine parts. This is called the three-view drawing (Fig.2.1.3).

2.2 Section view

A section view (Fig.2.1.4) shows a cross-section for a machine part.

2.3 Partial enlargement view

A partial enlargement view (Fig.2.1.5) is any view that is taken from a part of an existing view and scaled up for dimensioning and clarification.

Unit 2 Mechanical Drawing

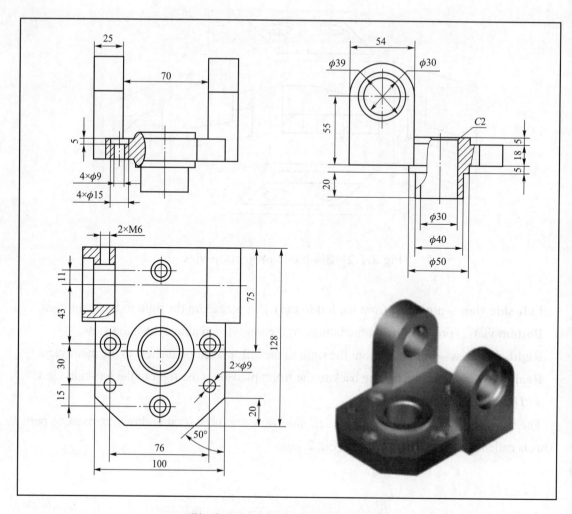

Fig.2.1.3 Three-view drawing

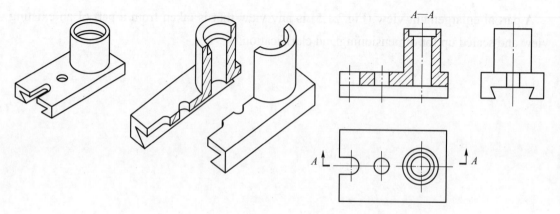

Fig.2.1.4 Section view

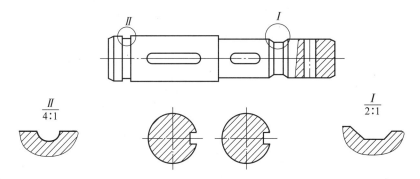

Fig.2.1.5 Partial enlargement view

New Words

2-1 单词

view [vju:]	n. [建筑学] 视图
item ['aitəm]	n. 项；条
scale [skeil]	n. 比例（尺）
	vt. 测量
projection [prə'dʒekʃn]	n. 投射；投影
visible ['vizəbl]	adj. 看得见的
invisible [in'vizəbl]	adj. 看不见的
orthographic [ˌɔːθə'ɡræfik]	adj. 正射；正字法的
regulation [ˌregju'leiʃn]	n. 规则；管理
standard ['stændəd]	n. 标准
plane [plein]	n. 平面；飞机
cube [kju:b]	n. 立方体；正方体
project ['prɔdʒekt]	vt. 投射；放映
rear [riə]	n. 后部；背面
section ['sekʃn]	n. 部分；剖面
partial ['pɑːʃəl]	adj. 部分的
enlargement [in'lɑːdʒmənt]	n. 放大；放大物
clarification [ˌklærəfi'keiʃn]	n. 清晰；说明

Phrases and Expressions

mechanical drawing	机械制图
projection view	投影视图
orthographic projection	正交投影法

Unit 2 Mechanical Drawing

main view   主视图
three-view drawing   三视图
section view   剖视图
partial enlargement view   局部放大图

Exercises

I. Read and judge.

(   ) 1. Mechanical drawing represents the physical shapes and sizes of the items they describe.

(   ) 2. Views must show both visible parts and invisible parts of a component.

(   ) 3. The method of drawing is orthographic projection.

(   ) 4. The left-side view is projected from the right to the left. It is placed on the left to the main view.

(   ) 5. A partial enlargement view is any view that is taken from a part of an existing view and scaled up for dimensioning and clarification.

II. Translate the following phrases.

English	Chinese
1. mechanical drawing	
2. main view	
3. three-view drawing	
4. partial enlargement view	
5.	投影视图
6.	基本视图
7.	仰视图
8.	剖视图

III. Fill in the blanks according to the text.

1. Mechanical drawing refers to using drawings to describe how a product is made or _____.

2. Views just show the _____ parts of a component; the _____ parts are drawn out only when necessary.

3. The main view is projected from _____.
The top view is projected from _____; placed _____.
The left-side view is projected from _____; placed _____.

The bottom view is projected from _____; placed _____.
The right-side view is projected from _____; placed _____.
The rear view is projected from _____; placed _____.

4. A section view shows a _____ for a machine part.

5. A partial enlargement view is any view that is taken from a part of an _____ and scaled up for _____.

IV. Fill in the blanks.

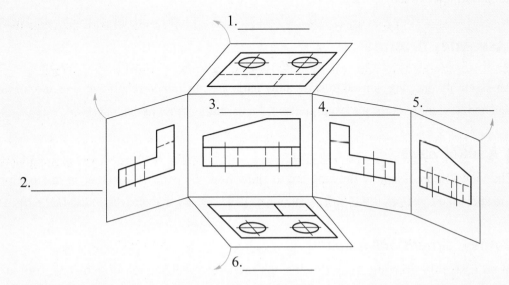

V. Translate the following sentences into Chinese.

1. Mechanical drawing refers to using drawings to describe how a product is made or assembled.

_____

2. The drawings of machine parts by projection are called views.

_____

3. Projection view is an orthographic projection of a machine part seen from the sides of front, top, right, etc.

_____

4. The main view, the top view and the left-side view are often used in drawing machine parts. This is called the three-view drawing.

_____

5. A section view shows a cross-section for a machine part.

_____

Lesson 2 Assembly drawing and detail drawing

There are two important types of mechanical drawings. One type is an assembly drawing, giving all information about a machine or a group of parts. The other one is a detail drawing, giving information about a machine part.

1. Assembly drawing

An assembly drawing is used to show how parts are positioned relative to one another in a design and how those parts are fixed in place. An assembly drawing includes:

1.1 A set of views

The views in an assembly drawing are to show how the parts fit together in the assembly and to show the function of the entire unit, not to describe the shape of the individual parts.

1.2 A few dimensions required

In an assembly drawing, such dimensions include: specification dimensions, mounting dimensions and other dimensions required.

1.3 Item numbers

Every part should be listed by the item numbers on the assembly drawing.

1.4 A list of the items and title block

All the assembly drawings have a list of the items placed above the title block or in a separate paper. The title block should be filled with the items such as the name of the machine, drawing number, weight, scale, etc.

1.5 Technical requirements

Technical requirements include all the information necessary for making, checking, mounting and maintaining the machine or component parts of a machine.

The following is an assembly drawing of a slide bearing (Fig.2.2.1). As shown in the title block, the slide bearing consists of eight kinds of parts.

① bearing housing
② lower bushing

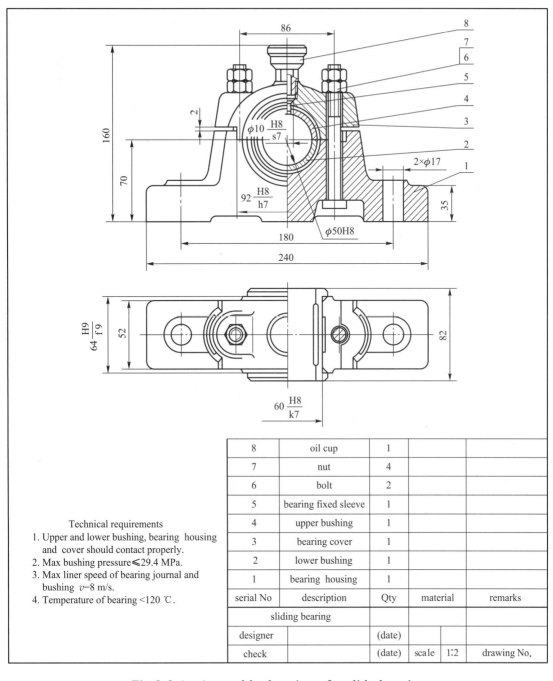

Fig.2.2.1　Assembly drawing of a slide bearing

③ bearing cover
④ upper bushing
⑤ bearing fixed sleeve
⑥ bolt
⑦ nut
⑧ oil cup

2. Detail drawing

A detail drawing (Fig.2.2.2) is used as an instruction for manufacture and inspection of a part. The detail drawing gives a complete and exact description of the form, dimensions and construction of a part. It includes:

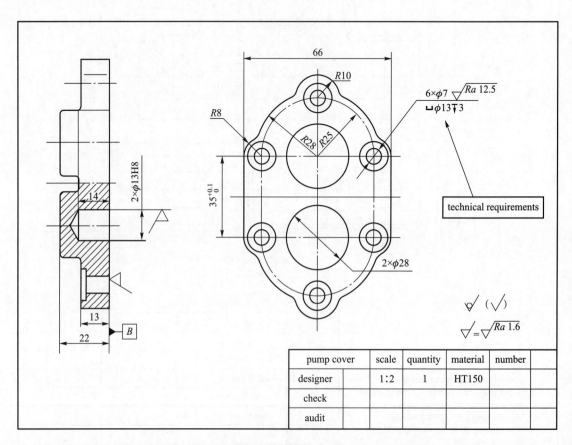

Fig.2.2.2　Detail drawing of a bearing block

2.1 A set of drawings

The drawings in a detail drawing show the internal and external shape of a part.

2.2 Overall dimensions

In a detail drawing, dimensions describe the size and the relative position of the parts. They should be correct, clear and reasonable.

2.3 Necessary technical requirements

Codes, symbols and notes are used to describe the necessary technical requirements in the process of manufacture, inspection and assembly.

2.4 Full contents of a title block

The title block on a detail drawing typically includes the part name, material, drawing number, date, scale and signature of responsible individual.

New Words

2-2 单词

detail ['di:teil]	n. 详述;(绘画等的)细部
entire [in'taiə]	adj. 全部的;整个的
individual [ˌindi'vidʒuəl]	adj. 个人的;个别的
dimension [di'menʃən]	n. 尺寸
specification [ˌspesifi'keiʃn]	n. 规格
bushing ['buʃiŋ]	n. 轴衬;套管
sleeve [sli:v]	n. [机]套筒;套管;袖子
instruction [in'strʌkʃn]	n. 使用说明;指令
inspection [in'spekʃn]	n. 检查;检验
description [di'skripʃn]	n. 描述
construction [kən'strʌkʃn]	n. 建造;构建
external [ik'stə:nl]	adj. 外面的
code [kəud]	n. 代码
symbol ['simbl]	n. 符号;标志
signature ['signətʃə]	n. 签名
responsible [ri'spɔnsəbl]	adj. 负有责任的;有责任的

Phrases and Expressions

assembly drawing	装配图
detail drawing	零件图
specification dimension	规格尺寸
mounting dimension	安装尺寸
technical requirements	技术要求
title block	标题栏

Unit 2 Mechanical Drawing

Exercises

I. Read and judge.

(   ) 1. There are two important types of mechanical drawings. They are an assembly drawing and a detail drawing.

(   ) 2. A detail drawing gives information about the whole machine.

(   ) 3. An assembly drawing includes a set of views, a few dimensions required, item numbers, a list of the items, title block and technical requirements.

(   ) 4. The title block should be filled with the items such as the name of the machine, drawing number, weight, scale, etc.

(   ) 5. An assembly drawing is used as an instruction for manufacture and inspection of a part.

II. Translate the following phrases.

English	Chinese
1. mounting dimension	
2. specification dimension	
3. external shape	
4. responsible individual	
5.	装配图
6.	零件图
7.	技术要求
8.	标题栏

III. Fill in the blanks according to the text.

1. There are two important types of mechanical drawings. One type is _____, giving all information about a machine or a group of parts. The other one is _____, giving information about a machine part.

2. An assembly drawing is used to show how parts are _____ relative to one another in a design and how those parts are _____ in place.

3. In an assembly drawing, such dimensions include: _____ dimensions, _____ dimensions and other dimensions required.

4. Technical requirements include all the information necessary for _____, _____, _____ and _____ the machine or component parts of a

machine.

5. A detail drawing is used as an instruction for _____ and _____ of a part.

IV. Translate the following sentences into Chinese.

1. An assembly drawing is used to show how parts are positioned relative to one another in a design and how those parts are fixed in place.
_____

2. Technical requirements include all the information necessary for making, checking, mounting and maintaining the machine or component parts of a machine.
_____

3. A detail drawing is used as an instruction for manufacture and inspection of a part. It gives a complete and exact description of the form, dimensions and construction of a part.
_____

4. The drawings in a detail drawing show the internal and external shape of a part.
_____

5. The title block on a detail drawing typically includes the part name, material, drawing number, date, scale and signature of responsible individual.
_____

Lesson 3　Introduction to AutoCAD

1. What is AutoCAD

Computer aided design (CAD) is the process of doing designs with the aid of computers. AutoCAD is a computer aided design software developed by Autodesk Inc. The AutoCAD software is a tool that can be used for design and drafting activities. The two-dimensional and three-dimensional models created in AutoCAD can be transferred to other computer programs for further analysis and testing.

2. Basic elements of AutoCAD user interface

Interface (Fig.2.3.1) items allow you to input data to and receive outputs from a computer system. AutoCAD interface can be divided into several elements, including: application menu, quick access toolbar, infocenter, ribbon tabs and panels, drawing area, cross cursor, command prompt area, status bar and so on.

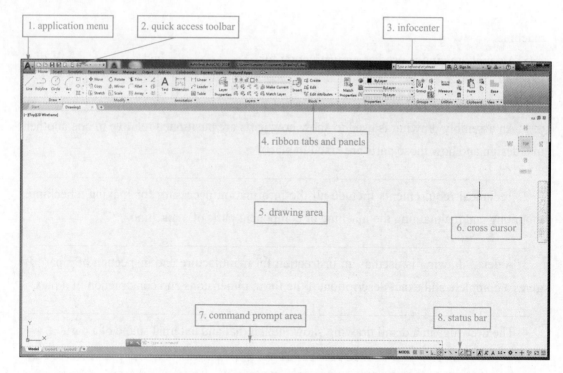

Fig.2.3.1　AutoCAD 2019 user interface

2.1 Application menu

The application menu is at the top of the main window and it contains commonly used file operations. This menu displays commands for creating a file, opening a file, saving a file, printing a file, closing a file, and other non-drawing tools.

2.2 Quick access toolbar

The quick access toolbar (Fig.2.3.2) is for a quick access to common commands like New, Open, Save, Undo and so on.

2.3 Infocenter

The infocenter (Fig.2.3.3) is for typing in keywords to search both online and offline resources and providing you with a list of related help topics.

Fig.2.3.2　Quick access toolbar

Fig.2.3.3　Infocenter

2.4 Ribbon tabs and panels

The ribbon (Fig.2.3.4) has most of the commands or tools used while working. It consists of two parts: tabs and panels.

Fig.2.3.4   Ribbon

- Tabs are on the top of ribbon. A series of tabs (Home, Insert, Manage, etc.) make up the ribbon and organize the tools into common groups.
- Panels contain different icons. For example, the Draw panel (Fig.2.3.5) contains icons for basic draw commands, such as Line, Polyline, Circle, Arc and so on.

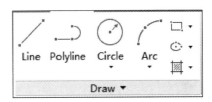

Fig.2.3.5   Draw panel

2.5 Drawing area

The drawing area (Fig.2.3.6) is the area where models and drawings are displayed.

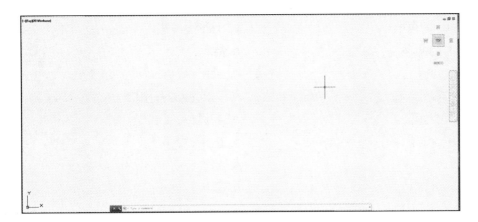

Fig.2.3.6   Drawing area

2.6 Cross cursor

The AutoCAD cross cursor is the primary means of pointing to and selecting objects or locations in the drawing area.

2.7 Command prompt area

The command prompt area (Fig.2.3.7) provides the status information for an operation and it is also the area for data input.

Fig.2.3.7　Command prompt area

2.8 Status bar

The status bar (Fig.2.3.8) shows the status of several commonly used display and construction options.

Fig.2.3.8　Status bar

2-3 单词

New Words

drafting [ˈdrɑːftiŋ]	n. 制图；绘图
transfer [trænsˈfəː]	vt. 转移；转化
interface [ˈintəfeis]	n. 界面
menu [ˈmenjuː]	n. 菜单
access [ˈækses]	n. 访问；入口；出口
toolbar [ˈtuːlbɑː]	n. 工具栏；工具条
ribbon [ˈribən]	n. 色带；彩带
tab [tæb]	n. 标签
panel [ˈpænl]	n. 面板
cursor [ˈkəːsə]	n. 光标
command [kəˈmɑːnd]	n. 命令；指挥
prompt [prɔmpt]	n. 提示符
status [ˈsteitəs]	n. 状态；地位
file [fail]	n. 文件
display [diˈsplei]	vt. 陈列；展览
create [kriˈeit]	vt. 创造
undo [ʌnˈduː]	vi. 撤销

resource [ri'sɔ:s]		n. 资源
icon ['aikɔn]		n. 图标
polyline ['pɔli:lain]		n. 多段线；折线
circle ['sə:kl]		n. 圆
arc [a:k]		n. 弧

Phrases and Expressions

application menu	应用菜单
quick access toolbar	快捷工具栏
infocenter	信息中心
cross cursor	十字光标
command prompt area	命令提示区
status bar	状态栏

Exercises

I. Read and judge.

(   ) 1. Computer aided design is the process of doing designs with the aid of computers.

(   ) 2. Only the two-dimensional models created in AutoCAD can be transferred to other computer programs for further analysis and testing.

(   ) 3. The application menu is at the bottom of the main window.

(   ) 4. The infocenter is for typing in keywords to search only online resources and providing you with a list of related help topics.

(   ) 5. The ribbon consists of two parts: tabs and panels.

II. Translate the following phrases.

English	Chinese
1. ribbon tabs and panels	
2. command prompt area	
3. infocenter	
4. status bar	
5.	绘图区
6.	应用菜单
7.	十字光标
8.	快捷工具栏

III. Fill in the blanks according to the text.

1. The AutoCAD software is a tool that can be used for _____ and _____ activities.

2. Interface items allow you to _____ data to and receive _____ from a computer system.

3. This menu displays commands for _____ a file, _____ a file, _____ a file, _____ a file, closing a file, and other non-drawing tools.

4. The ribbon consists of two parts: _____ and _____.

5. The AutoCAD cross cursor is the primary means of _____ to and _____ objects or locations in the drawing area.

IV. Fill in the basic elements of AutoCAD user interface.

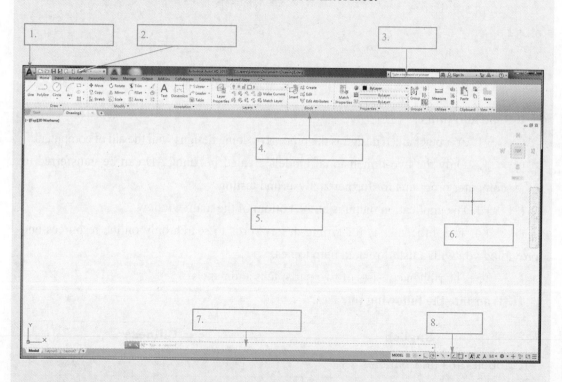

V. Translate the following sentences into Chinese.

1. Interface items allow you to input data to and receive outputs from a computer system.
_____

2. The quick access toolbar is for a quick access to common commands like New, Open, Save, Undo and so on.
_____

3. The ribbon has most of the commands or tools used while working.
_____

4. The AutoCAD cross cursor is the primary means of pointing to and selecting objects or locations in the drawing area.

5. The command prompt area provides the status information for an operation and it is also the area for data input.

Unit 3 Electronic Components and Circuits

In this unit, you will learn

- ◇ 1. Common electronic components;
- ◇ 2. Electric circuits;
- ◇ 3. Introduction to Multisim.

Lesson 1 Common electronic components

Electronic components (Fig.3.1.1) are basic electronic elements. They have two or more connecting leads or metallic pads. These components are connected together, usually by soldering to a printed circuit board, to create an electronic circuit with a particular function.

Fig.3.1.1 Common electronic components

1. Resistor

The resistor (Fig.3.1.2) is a two-terminal electrical component that resists electrical current. In electric circuits, resistors are used to reduce current flow, adjust signal levels, and divide voltages, etc. The opposition to the flow of current is termed resistance. The higher the value of resistance is, the lower the current will be.

2. Capacitor

The capacitor (Fig.3.1.3) is a device used to store charge in an electric circuit. A basic capacitor is made up of two conductors separated by an insulator or dielectric. Capacitor functions much like a battery, but charges and discharges much more efficiently (batteries, though, can store much more charge). The ability of a capacitor to store charge is called capacitance. The unit of

 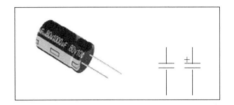

Fig.3.1.2 Resistor and its circuit symbol 　　Fig.3.1.3 Capacitor and its circuit symbols

capacitance is the farad (F).

3. Diode

The diode (Fig.3.1.4) is basically a one-way valve for electrical current. It lets the current flow in one direction (from positive to negative) and not in the other direction. Most diodes are similar to resistors in appearance but have a painted line on one end showing the direction of flow (white side is negative).

- *Light-emitting Diode (LED)*

A light-emitting diode (LED) (Fig.3.1.5) is an electronic light source. LEDs are based on the semiconductor diode. LEDs have many advantages compared to traditional light sources including lower energy consumption, longer lifetime, smaller size and faster switching.

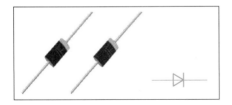 

Fig.3.1.4 Diodes and their circuit symbol　　Fig.3.1.5 LED and its circuit symbol

4. Transistor

A transistor (Fig.3.1.6) is a semiconductor device with at least three terminals for connection to a circuit. There are two types of standard transistors, NPN (negative—positive—negative)

Fig.3.1.6 Transistors and their circuit symbols

and PNP (positive—negative—positive), with different circuit symbols.

Transistors perform two basic functions. (1) It acts as a switch turning current on and off. (2) It acts as an amplifier. This makes an output signal that is a magnified version of the input signal.

3-1 单词

New Words

pad [pæd]	n. 垫片;焊盘
solder [ˈsəuldə]	vt. & vi. 焊接;焊锡
resistor [riˈzistə]	n. 电阻器
terminal [ˈtə:minl]	n. 终端;(电路的)端子
current [ˈkʌrənt]	n. 电流
voltage [ˈvəultidʒ]	n. 电压
capacitor [kəˈpæsitə]	n. 电容器
charge [tʃɑ:dʒ]	n. 电荷
	vi. 充电
conductor [kənˈdʌktə]	n. [电]导体
insulator [ˈinsjuleitə]	n. 绝缘体
dielectric [ˌdaiiˈlektrik]	n. 电介质;绝缘体
battery [ˈbætri]	n. 电池;蓄电池
discharge [disˈtʃɑ:dʒ]	vi. 放电;释放
capacitance [kəˈpæsitəns]	n. 电容;电流容量
farad [ˈfæræd]	n. 法拉(电容单位)
diode [ˈdaiəud]	n. 二极管
positive [ˈpɔzətiv]	adj. 正的;积极的
negative [ˈnegətiv]	adj. 负的;消极的
emit [iˈmit]	vt. 发出;发射
semiconductor [ˌsemikənˈdʌktə]	n. [物]半导体
switch [switʃ]	n. 开关;转换
	vt. & vi. 转换
transistor [trænˈzistə]	n. 晶体管;半导体收音机
amplifier [ˈæmplifaiə]	n. 放大器
magnify [ˈmægnifai]	vt. 放大

Lesson 1 Common electronic components

Phrases and Expressions

light-emitting diode (LED)   发光二极管

Exercises

I. Read and judge.

(   ) 1. Electronic components have two or more connecting leads or metallic pads.
(   ) 2. The resistor is a three-terminal electrical component that resists electrical current.
(   ) 3. The capacitor is a device used to store charge in an electric circuit.
(   ) 4. LEDs have many advantages including lower energy consumption, longer lifetime, smaller size and faster switching.
(   ) 5. A transistor is a conductor device with at least three terminals for connection to a circuit.

II. Translate the following phrases.

English	Chinese
1. electronic circuit	
2. insulator	
3. light-emitting diode	
4.	导体
5.	充电
6.	半导体

III. Fill in the blanks.

1. The resistor is a two-terminal electrical component that resists electrical _____. The opposition to the flow of current is termed _____. The higher the value of resistance is, the lower the _____ will be.

2. The _____ is a device used to store charge in an electric circuit. A basic capacitor is made up of two conductors separated by an _____ or dielectric. Capacitor functions much like a _____, but charges and _____ much more efficiently. The ability of a capacitor to store charge is called _____. The unit of capacitance is the _____.

3. A _____ (LED) is an electronic light source. LEDs are based on the _____ diode.

4. There are two types of standard transistors, _____ (negative—positive—negative) and _____ (positive—negative—positive).

5. Transistors perform two basic functions. (1) It acts as a _____ turning current on and off. (2) It acts as an _____.

IV. Give the English and Chinese names of the components.

1. ![] 2. ![] 3. ![] 4. ![] 5. ![] ![]

English name _____ _____ _____ _____ _____
Chinese name _____ _____ _____ _____ _____

V. Translate the following sentences into Chinese.

1. The resistor is a two-terminal electrical component that resists electrical current.

2. The capacitor is a device used to store charge in an electric circuit.

3. The diode is basically a one-way valve for electrical current.

4. A light-emitting diode (LED) is an electronic light source.

5. A transistor is a semiconductor device with at least three terminals for connection to a circuit.

Lesson 2 Electric circuits

An electric circuit (Fig.3.2.1) is a path in which electric current flows.

A simple electric circuit consists of three main elements: an electrical source, an electrical

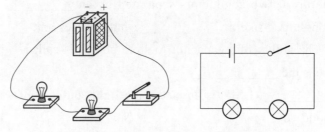

Fig.3.2.1 Simple electric circuit

load and an electrical wire.

The electrical source can be any source of electrical energy. In practice, it may be a battery, an electrical generator, or some sort of electronic power supply.

The electrical load is any device or circuit powered by electricity. It can be as simple as a light bulb or as complex as a modern high-speed computer.

The electrical wire conducts electricity from the source to the load and back again. The wire is commonly made of copper or aluminum, and is usually insulated to prevent shocks and short circuits.

1. Power source

Circuits use two forms of electrical power: alternating current (AC) and direct current (DC) (Fig.3.2.2). AC often powers large appliances and motors and is generated by power stations. DC powers battery operated vehicles, other machines and electronics.

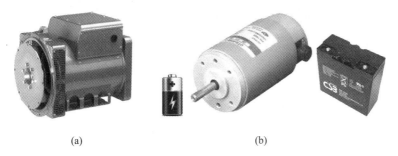

Fig.3.2.2 AC generator and DC power supply

2. Voltage

Voltage is the difference in electrical potential between two points in a circuit. Voltage is represented by the letter U and is measured with a voltmeter. The unit of voltage is volt (V) which was named after the Italian physicist Alessandro Volta (Fig.3.2.3) who made the first chemical battery.

Fig.3.2.3 Alessandro Volta and voltmeter

3. Current

The current is a flow of electric charge. In electric circuits, electric charge is often carried by moving electrons in a wire.

The symbol for current is *I*, which originates from the French phrase intensité de courant, meaning current intensity. Current intensity is often referred to simply as current. The symbol *I* was first used by French physicist A. Ampere (Fig.3.2.4). The unit of electric current, ampere (A), was named after him. Electric current is measured with a device called an ammeter.

Electric currents cause Joule heat, which creates light in light bulbs. They also create magnetic fields, which are used in motors, inductors and generators.

Fig.3.2.4   A. Ampere and ammeter

4. Resistance

Resistors are used to increase resistance in the circuit, so it slows down the current. The symbol for resistance is *R*. The unit of electrical resistance is ohm (Ω) which was named after German physicist Georg Simon Ohm (Fig.3.2.5).

Resistance determines how much current flows through a component. High resistance allows a small amount of current to flow. Low resistance allows a large amount of current to flow. Ohm's Law defines the relationships between voltage (*U*), current (*I*) and resistance (*R*).

Fig.3.2.5   Georg Simon Ohm

Ohm's Law: The current (*I*) through a conductor between two points is directly proportional to the potential difference or voltage (*U*) across the two points, and inversely proportional to the resistance (*R*) between them. This is represented by the formula:

$$I=\frac{U}{R}$$

Lesson 2 Electric circuits

3-2 单词

New Words

source [sɔ:s]	n. 根源；本源
load [ləud]	n. 负载；负荷
generator [ˈdʒenəreitə]	n. 发电机
bulb [bʌlb]	n. 灯泡
complex [ˈkɔmpleks]	adj. 复杂的
insulate [ˈinsjuleit]	vt. 使隔离；使绝缘
alternate [ɔ:lˈtə:nət]	vt. & vi. 交替
appliance [əˈplaiəns]	n. 装置；家用电器
potential [pəˈtenʃl]	n. [物]电位；势
voltmeter [ˈvəultmi:tə]	n. 电压表
volt [vəult]	n. 伏特；电压
physicist [ˈfizisist]	n. 物理学家
electron [iˈlektrɔn]	n. 电子
originate [əˈridʒineit]	vt. 创始；开始
intensity [inˈtensəti]	n. 强度
ampere [ˈæmpeə]	n. [电]安培
ammeter [ˈæmi:tə]	n. 电流表
inductor [inˈdʌktə]	n. 电感器；感应器
ohm [əum]	n. 欧姆
relationship [riˈleiʃnʃip]	n. 关系
directly [dəˈrektli]	adv. 直接地；正好地
proportional [prəˈpɔ:ʃənəl]	adj. 比例的；成比例的
inversely [ˌinˈvə:sli]	adv. 相反地
formula [ˈfɔ:mjələ]	n. 公式

Phrases and Expressions

short circuit	短路
alternating current (AC)	交流电
direct current (DC)	直流电
power station	发电厂；发电装置
electric charge	电荷
Joule heat	焦耳热

directly proportional 正比例的
inversely proportional 反比例的

Exercises

I. Read and judge.

(　　) 1. The electrical wire conducts electricity from the load to the source and back again.
(　　) 2. Volt (V) was named after the Italian physicist Alessandro Volta.
(　　) 3. Electric current is measured with a device called an ammeter.
(　　) 4. High resistance allows a large amount of current to flow.
(　　) 5. Ohm's Law defines the relationships between voltage (U), current (I) and resistance (R).

II. Translate the following phrases.

English	Chinese
1. Ohm's Law	
2. alternating current	
3. electric charge	
4. Joule heat	
5.	短路
6.	直流电
7.	电源
8.	电流强度

III. Fill in the blanks.

1. A simple electric circuit consists of three main elements: an electrical _____, an electrical _____ and an electrical _____.

2. The electrical source can be any source of electrical energy. In practice, it may be a _____, an electrical _____, or some sort of electronic _____ supply.

3. The electrical load is any device or _____ powered by electricity.

4. Circuits use two forms of electrical power: _____ and _____.

5. _____ determines how much current flows through a component.

6. The current through a conductor between two points is _____ to the potential difference or voltage across the two points, and _____ to the resistance

between them.

IV. Fill in the blanks by choosing the words from the box.

R   U   I   volt   ohm   ampere   voltmeter   ammeter

1. Voltage is represented by the letter _____ and is measured with a _____. The unit of voltage is _____ (V) which was named after the Italian physicist Alessandro Volta who made the first chemical battery.

2. The symbol for current is _____. The unit of electric current, _____ (A), was named after A. Ampere. Electric current is measured with a device called an _____.

3. The symbol for resistance is _____. The unit of electrical resistance is _____ (Ω) which was named after German physicist Georg Simon Ohm.

V. Translate the following sentences into Chinese.

1. An electrical circuit is a path in which electric current flows.

2. Voltage is the difference in electrical potential between two points in a circuit.

3. The current is a flow of electric charge.

4. Electric currents cause Joule heat, which creates light in light bulbs.

5. High resistance allows a small amount of current to flow.

Lesson 3   Introduction to Multisim

3-3 课文

1. What is Multisim

The Multisim software is a program that acts as a virtual electronics laboratory. You can use the Multisim program not only to create electronic circuit on your computer, but also to simulate ("run") the circuits and use virtual laboratory instruments to make electronic measurements.

Multisim is widely used for circuits education, electronic schematic design, etc.

2. Basic user interface of Multisim

Once the Multisim program starts, you will see the user interface shown as follows (Fig.3.3.1).

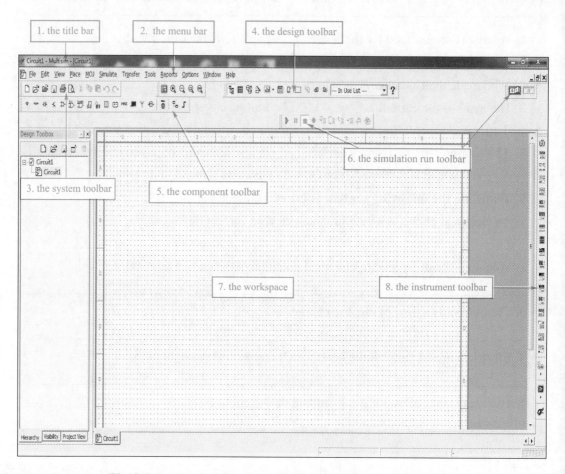

Fig.3.3.1　Basic elements of user interface of Multisim

2.1 Title bar

The title bar (Fig.3.3.2) is at the very top of the screen and is common to all Windows applications. The left side of the bar contains information about the application and the file, while the right side of the bar contains controls to minimize, maximize and close the application.

Fig.3.3.2　Title bar

2.2 Menu bar

The menu bar (Fig.3.3.3) contains standard Windows menus (such as File, Edit, View and Help) and menus that are specific to Multisim [such as Place, MCU (microcontroller unit), Simulate, etc].

Fig.3.3.3 Menu bar

2.3 System toolbar

The system toolbar (Fig.3.3.4) contains tools for performing common operations from the File and Edit menus, such as creating a new file, printing a file, etc.

Fig.3.3.4 System toolbar

2.4 Design toolbar

The design toolbar (Fig.3.3.5) contains tools to access various Multisim features and information about the current circuit.

Fig.3.3.5 Design toolbar

2.5 Component toolbar

The component toolbar (Fig.3.3.6) contains tools that let you access various components to create and analyze circuits.

Fig.3.3.6 Component toolbar

2.6 Simulation run toolbar

The simulation run toolbar (Fig.3.3.7) contains tools that let you start, stop a simulation, etc.

Fig.3.3.7 Simulation run toolbar

2.7 Instrument toolbar

The instrument toolbar (Fig.3.3.8) provides instruments with which you can measure and evaluate the operation of a circuit. Some instruments, like the multimeter, are real devices used to analyze real-world circuits. Other instruments, like the logic converter, exist only within the Multisim application and are convention tools for you to simulate, analyze and debug your circuit designs.

Fig.3.3.8　Instrument toolbar

3-3 单词

New Words

laboratory [ləˈbɔrətri]	n. 实验室
simulate [ˌsimjuˈleit]	n. 模仿；仿真
instrument [ˈinstrəmənt]	n. 仪器；器械
schematic [skiːˈmætik]	n. 电路原理图
minimize [ˈminimaiz]	vt. 最小化
maximize [ˈmæksimaiz]	vt. 最大化
analyze [ˈænəlaiz]	vt. 分析
simulation [ˌsimjuˈleiʃn]	n. 模仿；仿真
evaluate [iˈvæljueit]	vt. 求……的值；计值
multimeter [ˈmʌltimiːtə]	n. 多用表（又称万用表）
logic [ˈlɔdʒik]	n. 逻辑
converter [kənˈvəːtə]	n. 变换器；转换器
debug [ˌdiːˈbʌg]	vt. 调试；排除故障

Phrases and Expressions

title bar	标题栏
menu bar	菜单栏
microcontroller unit (MCU)	微控制器
system toolbar	系统工具栏
design toolbar	设计工具栏

English	Chinese
component toolbar	元件工具栏
simulation run toolbar	仿真运行工具栏
instrument toolbar	仪器仪表工具栏
logic converter	逻辑转换器

Exercises

I. Read and judge.

(　　) 1. The Multisim software is a program that acts as a virtual electronics laboratory.

(　　) 2. The Multisim program can not only create electronic circuit on your computer, but also simulate the circuits.

(　　) 3. The menu bar is the same as Windows menus.

(　　) 4. The system toolbar contains tools for performing common operations from the File and Edit menus.

(　　) 5. The design toolbar contains tools that let you access various components to create and analyze circuits.

II. Translate the following phrases.

English	Chinese
1. design toolbar	
2. simulation run toolbar	
3. microcontroller unit	
4. logic converter	
5.	元件工具栏
6.	标题栏
7.	仪器仪表工具栏
8.	系统工具栏

III. Fill in the blanks according to the text.

1. The Multisim software is a program that acts as a _____. You can use the Multisim program not only to create electronic circuit on your computer, but also to _____ ("run") the circuits.

2. _____ contains tools to access various Multisim features and information about the current circuit.

3. _____ contains tools that let you access various components to create and

_____ circuits.

4. The simulation run toolbar contains tools that let you start, stop a _____, etc.

5. The instrument toolbar provides instrument with which you can measure and _____ the operation of a circuit. Other instruments, like the _____, exist only within the Multisim application and are _____ for you to simulate, analyze and _____ your circuit designs.

IV. Translate the following sentences into Chinese.

1. The Multisim software is a program that acts as a virtual electronics laboratory.
_____

2. The design toolbar contains tools to access various Multisim features and information about the current circuit.
_____

3. The component toolbar contains tools that let you access various components to create and analyze circuits.
_____

4. The simulation run toolbar contains tools that let you start, stop a simulation, etc.
_____

5. The instrument toolbar provides instruments with which you can measure and evaluate the operation of a circuit.
_____

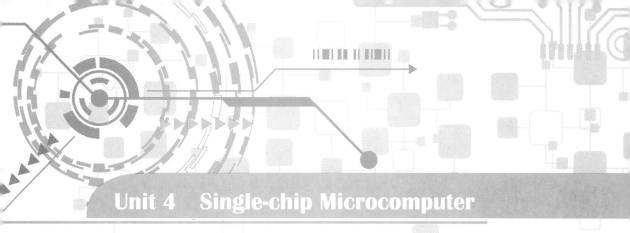

Unit 4　Single-chip Microcomputer

In this unit, you will learn

◇ 1. Introduction to single-chip microcomputer;
◇ 2. Applications of single-chip microcomputer;
◇ 3. MCS-51™ instruction set summary.

Lesson 1　Introduction to single-chip microcomputer

1. What is single-chip microcomputer

Single-chip microcomputer (SCM) is a complete computer system integrated on a single chip (Fig.4.1.1). It is also known as microcontroller unit (MCU). Micro suggests that the device is small, and controller tells you that the device might be used to control objects, processes or events.

Fig.4.1.1　Single-chip microcomputer

2. History of SCM

In 1975, the Altair 8800 computer (Fig.4.1.2) was the first microcomputer that hobbyists could build and program themselves. The basic Altair included no keyboard, video display, disk drives or other elements we now think of as essential elements of a personal computer.

Fig.4.1.2　Altair 8800 computer

It was programmed by flipping toggle switches on the front panel. Standard random access memory (RAM) was 256 bytes. A small company called Microsoft offered a version of the BASIC programming language for it.

The computer world has changed a lot since the introduction of the Altair. Microsoft has become an enormous software publisher. Building a personal computer now only requires installing assembled boards and other components in an enclosure. A personal computer (PC) like IBM's PC is general-purpose machine and you can use it for many applications—word processing, spreadsheets, etc.

But along with cheap and powerful PCs, small and customized computers for specific uses have developed. Each of these small computers is dedicated to one task, or a set of closely related tasks. At the core of many of these specialized computers is a microcontroller.

3. Basic system of SCM

SCM (Fig.4.1.3) has almost all the components of a computer, including the processor, the memory system and the input and output (I/O) system.

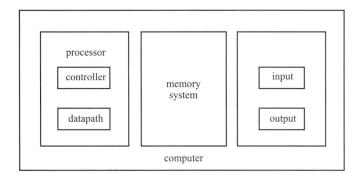

Fig.4.1.3　Basic system of SCM

The processor or central processing unit (CPU) contains both datapath and controller.

The memory system contains kinds of memory components to support the operation of the computer. Typical memory systems aboard SCM contain random access memory (RAM), read only memory (ROM) and electrically erasable programmable read only memory (EEPROM).

The input and output (I/O) system of SCM usually consists of a series of ports.

4. Peripheral modules of SCM

Most microcomputers contain a number of hardware modules. In the past, many of these were designed as separate chips in a microcomputer system. Integrating them into a single microcontroller chip allows for greater functionality in a single chip and saves space.

Typical devices are:
- Timer module;
- Serial I/O module;
- Analogue to digital converter module.

4-1 单词

New Words

chip [tʃip]	n. 芯片；碎片
microcontroller [maikrəukɔntˈrəulə]	n. 微控制器
hobbyist [ˈhɔbiist]	n. 业余爱好者
flip [flip]	vt. 按(开关)；轻弹，轻击
toggle [ˈtɔgl]	n. 转换键；切换键
byte [bait]	n. 字节
spreadsheet [ˈspredʃi:t]	n. 电子表格程序
processor [ˈprəusesə]	n. (计算机的)中央处理器
datapath [ˈdeitəˈpɑ:θ]	n. 通路
erasable [iˈreisəbl]	adj. 可消除的；可擦除的
port [pɔ:t]	n. 端口；港口
peripheral [pəˈrifərəl]	adj. 外围的；次要的
module [ˈmɔdju:l]	n. 模块；组件
analogue [ˈænəlɔg]	adj. 模拟计算机的

Phrases and Expressions

single-chip microcomputer (SCM)	单片机
word processing	字处理
central processing unit (CPU)	中央处理器
random access memory (RAM)	随机存取存储器
read only memory (ROM)	只读存储器
electrically erasable programmable read only memory (EEPROM)	电可擦只读存储器
timer module	定时模块
serial I/O module	串行输入/输出模块
analogue to digital converter module	模数转换模块

Lesson 1 Introduction to single-chip microcomputer

Exercises

I. Read and judge.
(   ) 1. Single-chip microcomputer (SCM) is also known as microcontroller unit (MCU).
(   ) 2. In 1975, the Altair 8800 computer was the first microcomputer that scientists could build and program themselves.
(   ) 3. SCM has almost all the components of a computer.
(   ) 4. The central Processing unit (CPU) contains both datapath and controller.
(   ) 5. Most microcomputers contain a number of software modules.

II. Translate the following phrases.

English	Chinese
1. microcontroller unit	
2. random access memory	
3. electrically erasable programmable read only memory	
4. peripheral modules	
5.	单片机
6.	中央处理器
7.	只读存储器
8.	定时模块

III. Fill in the blanks.
1. _____ (SCM) is a complete computer system integrated on a single chip. It is also known as _____ (MCU). _____ suggests that the device is small, and _____ tells you that the device might be used to control objects, processes or events.

2. SCM has almost all the components of a computer, including _____, _____ and _____.

3. The processor or _____ (CPU) contains both _____ and _____.

4. Typical memory systems aboard SCM contain _____ (RAM), _____ (ROM) and _____ (EEPROM).

5. Most microcomputers contain a number of hardware modules. Typical devices are: _____, _____ and _____.

IV. Fill the blanks.

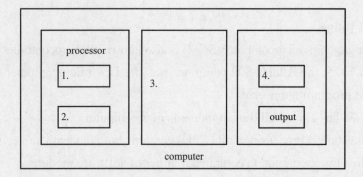

V. Translate the following sentences into Chinese.

1. Single-Chip Microcomputer (SCM) is a complete computer system integrated on a single chip. It is also known as microcontroller unit (MCU).

_____

2. SCM has almost all the components of a computer, including the processor, the memory system and the input and output (I/O) system.

_____

3. The processor or central processing unit (CPU) contains both datapath and controller.

_____

4. The input and output (I/O) system of SCM usually consists of a series of ports.

_____

4-2 课文

Lesson 2 Applications of single-chip microcomputer

Single-chip microcomputer has a small size, low power consumption and controlling function. Since it is easy to use, it is widely used in household appliances, industrial control, intelligent instruments, the field of automotive equipment, and the field of computer networks and communications, etc.

1. Applications in household appliances

The household appliances basically use single-chip microcomputer, such as video recorders, cameras, fully automatic washing machines, refrigerators, air conditioners, color TVs, program-controlled toys (Fig.4.2.1), etc. The electronic weighing equipment is also controlled by a single-chip microcomputer.

2. Applications in industrial control

In the industrial area, a single-chip microcomputer can constitute a variety of control systems, data acquisition systems (Fig.4.2.2), such as factory assembly line of an intelligent control.

Fig.4.2.1 Program-controlled toy

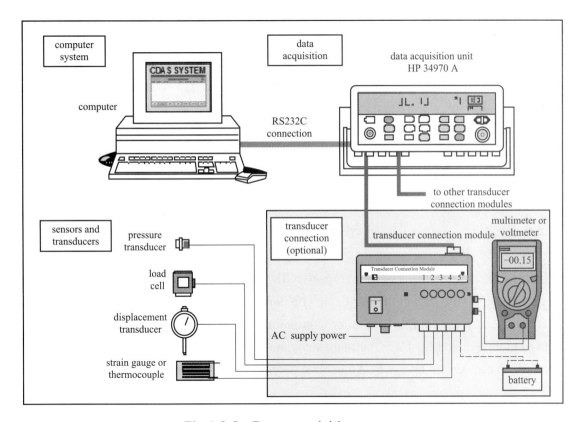

Fig.4.2.2 Data acquisition system

3. Applications in intelligent instruments

Single-chip microcomputer is widely used in instruments. Combining different types of sensors, it can implement measurement of voltage, power, frequency, temperature (Fig.4.2.3), flow, speed, thickness, angle, length, pressure, etc. With the digital instruments, the intelligence, miniaturization and functionality have been realized. SCM is more powerful than electronic or digital circuits, such as precision measuring equipment (power meter, analytical instrument).

4. Applications in the field of automotive equipment

Single-chip microcomputer is widely used in automotive electronics, such as a vehicle engine controller, GPS (global position system) navigation system (Fig.4.2.4), ABS (anti-lock braking system), etc.

Fig.4.2.3 Temperature meter   Fig.4.2.4 GPS navigation system

5. Applications in the field of computer networks and communications

Generally, single-chip microcomputer with modern communication interfaces can communicate easily with computer data. The communications equipment can achieve the intelligent control by single-chip microcomputer from program-controlled switchboards, trunked mobile radio (Fig.4.2.5), building automated communications call systems to the daily work which can be seen everywhere in mobile phones, walkie-talkies (Fig.4.2.6), etc.

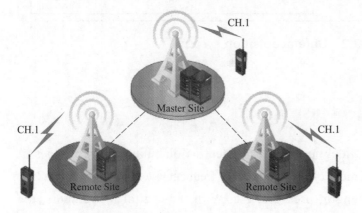 

Fig.4.2.5 Trunked mobile radio   Fig.4.2.6 Walkie-talkie

In fact, almost every piece of electronic and mechanical products used in modern human life has a single-chip microcomputer.

New Words

4-2 单词

refrigerator [riˈfridʒəreitə]　　　　　　n. 冰箱
frequency [ˈfrikwənsi]　　　　　　　　n. 频率
temperature [ˈtemprətʃə]　　　　　　　n. 温度
intelligence [inˈtelidʒəns]　　　　　　　n. 智能
miniaturization [ˌminətʃəraiˈzeiʃn]　　　n. 小型化；微型化
analytical [ˌænəˈlitikl]　　　　　　　　adj. 分析的
switchboard [ˈswitʃbɔːd]　　　　　　　n. 交换台；总机
walkie-talkie [ˌwɔːki ˈtɔːki]　　　　　　n. 手持式对讲机；步话机

Phrases and Expressions

household appliances　　　　　　　　家用电器
video recorder　　　　　　　　　　　录像机
air conditioner　　　　　　　　　　　空气调节机；空调设备
data acquisition system　　　　　　　数据采集系统
factory assembly line　　　　　　　　工厂装配线
global position system (GPS)　　　　　全球定位系统
GPS navigation system　　　　　　　　GPS 导航系统
anti-lock braking system (ABS)　　　　防抱死制动系统
program-controlled switchboard　　　　程控交换机
trunked mobile radio　　　　　　　　集群移动通信
building automated communications call system　　楼宇自动通信呼叫系统

Exercises

I. Read and judge.

(　　) 1. Single-chip microcomputer has a small size, but high power consumption.

(　　) 2. The electronic weighing equipment can't be controlled by single-chip microcomputer.

(　　) 3. Single-chip microcomputer is widely used in instruments.

(　　) 4. SCM is less powerful than electronic or digital circuits.

(　　) 5. The communication equipment can achieve an intelligent control by any single-chip microcomputer.

II. Translate the following phrases.

English	Chinese
1. intelligent instruments	
2. program-controlled switchboard	
3. global position system	
4. factory assembly line	
5.	GPS 导航系统
6.	工业控制
7.	防抱死制动系统
8.	手持式对讲机

III. Fill in the blanks according to the text.

1. Single-chip microcomputer has a small size, _____ and controlling function. Since it is easy to use, it is widely used in _____, industrial control, intelligent instruments, the field of _____, and the field of _____, etc.

2. In the industrial areas, a single-chip microcomputer can constitute a variety of control systems, data acquisition systems, such as _____.

3. Combining different types of sensors, it can implement measurement of _____, power, _____, _____, flow, speed, thickness, angle, length, pressure, etc.

IV. Translate the phrases into Chinese, then match with the pictures.

1. temperature meter　　_____ (　　)

A.

B.

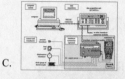
C.

D.

E.

F.

2. walkie-talkie _____ ( )
3. program-controlled toy _____ ( )
4. GPS navigation system _____ ( )
5. data acquisition system _____ ( )
6. trunked mobile radio _____ ( )

V. Translate the following sentences into Chinese.

1. The household appliances basically use single-chip microcomputer.

_____

2. In the industrial area, a single-chip microcomputer can constitute a variety of control systems, data acquisition systems.

_____

3. Generally, single-chip microcomputer with modern communication interfaces can communicate easily with computer data.

_____

4. In fact, almost every piece of electronic and mechanical products used in modern human life has a single-chip microcomputer.

_____

Lesson 3   MCS-51™ instruction set summary

4-3 课文

1. What is MCS-51™ series single-chip microcomputer

One of the most successful SCM in early single-chips is Intel's 8031, which is simple and reliable. A single-chip microcomputer system MCS-51 series has been developed since then in Intel's 8031. Based on the single-chip microcomputer system, MCS-51 series (Fig.4.3.1) has been widely used until now.

2. Structure of SCM instructions

Fig.4.3.1   MCS-51 series

Usually there are two parts in each instruction of SCM (Fig.4.3.2).

The first part of each instruction is called mnemonics referring to the operation that an instruction performs (copy, addition, logic operation, etc.). Mnemonics are abbreviations of the name of the operations.

For example:
- DEC—decrement;
- INC—increment;
- CLR—clear;
- DIV—divide;
- MUL—multiply;
- SUBB — subtract with borrow.

The other part of the instruction is called operand which defines data being processed by instructions.

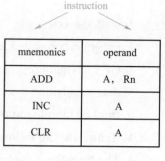

Fig.4.3.2 Structure of SCM instructions

- A—accumulator;
- Rn — one of working registers (R0~R7) in the currently active RAM memory bank;
- #data —an 8-bit constant included in instruction;
- addr16 —a 16-bit address.

3. Types of instructions

Depending on performing operations, all instructions are divided in several groups:

3.1 Arithmetic instructions
- ADD A, Rn —adds the register Rn to the accumulator;
- DEC A —decrements the accumulator by 1;
- DIV AB —divides the accumulator by the register B;
- INC A—Increments the accumulator by 1;
- MUL AB—multiplies A and B.

3.2 Branch instructions
JNZ rel—jump if accumulator is not zero.

3.3 Logic instructions
CLR A—clears the accumulator.

3.4 Data transfer instructions
PUSH direct—pushes the direct byte onto the stack.

3.5 Bit-oriented instructions
Bit: direct addressed bit in RAM;
CLR bit—clears the direct bit.

New Words

4-3 单词

mnemonic [niˈmɔnik]	n. 助记符;记忆术
abbreviation [əˌbriviˈeiʃən]	n. 省略;缩略词
operand [ˈɔpərænd]	n. 操作数;运算数;运算对象
decrement [ˈdekrimənt]	n. 减量;衰减量;减一
increment [ˈinkrəmənt]	n. 增量;增加;增值;增长
multiply [ˈmʌltiplai]	vt. & vi. 乘;(使)相乘
subtract [səbˈtrækt]	vt. 减去;扣除
accumulator [əˈkju:mjəleitə]	n. 累加器
register [ˈredʒistə]	n. 寄存器;登记;注册账户
constant [ˈkɔnstənt]	n. 常数;常量
stack [stæk]	n. 堆栈

Phrases and Expressions

arithmetic instruction	算术指令
branch instruction	控制转移指令
data transfer instruction	数据转移指令
logic instruction	逻辑指令
bit-oriented instruction	位操作指令

Exercises

I. Read and judge.

(　　) 1. One of the most successful SCM is Intel's 8031, which is simple and reliable.

(　　) 2. MCS-51 series is not widely used until now.

(　　) 3. Mnemonics are the names of the operations.

(　　) 4. Operand defines data being processed by instructions.

(　　) 5. SUBB is the abbreviation of subtract.

II. Translate the following phrases.

English	Chinese
1. arithmetic instructions	
2. branch instructions	

	continued
English	**Chinese**
3. data transfer instructions	
4. bit-oriented instructions	
5.	助记符
6.	操作数
7.	累加器
8.	寄存器

III. Give the whole words of the following abbreviations.

1. DEC _____ 2. INC _____
3. CLR _____ 4. DIV _____
5. MUL _____ 6. SUBB _____

IV. Explain the following instructions and translate them into Chinese.

1. DEC A _____
2. DIV AB _____
3. INC A _____
4. MUL AB _____
5. JNZ rel _____
6. CLR A _____

Unit 5 Programmable Logic Controller (PLC)

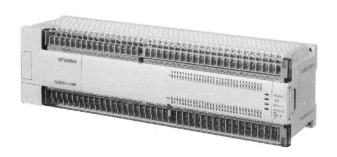

In this unit, you will learn

◇ 1. Introduction to PLC;
◇ 2. PLC operations;
◇ 3. Troubleshooting of PLC.

Lesson 1 Introduction to PLC

1. What is PLC

A programmable logic controller (PLC) (Fig.5.1.1) is a special form of microcontroller. It uses programmable memory to store instructions and implement functions such as logic, sequencing, timing, counting, and arithmetic in order to control machines and processes. The term logic is used because programming is mainly concerned with implementing logic and switching operations.

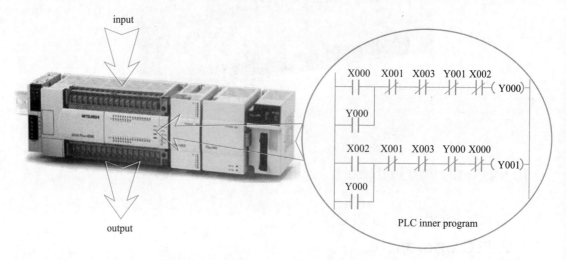

Fig.5.1.1 Programmable logic controller

2. History of PLCs

The early programmable logic controller was designed and developed by Modicon Company as a relay replacer (Fig.5.1.2) for General Motors (GM) and Landis. The new system increased the functionality of the controls while reducing the cabinet space that housed the logic.

The first PLC, model 084, was invented by Dick Morley with his Modicon Team (Fig.5.1.3) in 1969.

The first successful commercial PLC, model 184, was introduced in 1973 and was designed by Michael Greenberg.

Fig.5.1.2 Complicated relay replacer used in 1960's

Fig.5.1.3 Modicon team with the 084 PLC

3. Basic components in a typical PLC system (Fig.5.1.4)

Typically PLC (Fig.5.1.5) has the basic components of processor unit, memory, power supply unit, input and output sections, communications and programming device.

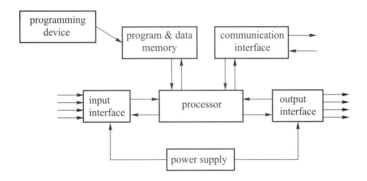

Fig.5.1.4 Basic arrangements of PLC

Unit 5 Programmable Logic Controller (PLC)

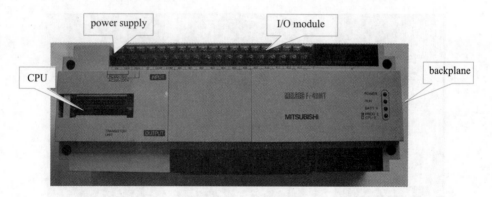

Fig.5.1.5   PLC

The processor unit or central processing unit (CPU) is the unit containing the microprocessor. This unit interprets the input signals and carries out control actions according to the program stored in its memory, communicating the decisions as action signals to the outputs.

The power supply unit is needed to convert the mains AC voltage to the low DC voltage (5 V) necessary for the processor and the circuits in the input and output interface modules.

The programming device (Fig.5.1.6) is used to enter the required program into the memory of the processor. The program is developed in the device and then transferred to the memory unit of the PLC.

Fig.5.1.6   Programming devices

New Words

implement [ˈimplimənt]　　　　　　　vt. 实施；执行
sequence [ˈsikwəns]　　　　　　　　vt. 按顺序排好
relay [ˈri:lei]　　　　　　　　　　　n. 继电器
replacer [riˈpleisə]　　　　　　　　　n. 代用品

cabinet ['kæbinət]　　　　　　　　　　　n. 柜；橱
commercial [kə'mə:ʃl]　　　　　　　　　adj. 商业的
interpret [in'tə:prit]　　　　　　　　　　vt. 解释；理解；翻译；译码
convert [kən'və:t]　　　　　　　　　　 vt. （使）转变；转换；变换
mains [meinz]　　　　　　　　　　　　n. 总输电线；干线；电源

Phrases and Expressions

programmable logic controller (PLC)　　可编程逻辑控制器
Modicon Company　　　　　　　　　　莫迪康公司
General Motors (GM)　　　　　　　　　通用汽车（公司）
Landis　　　　　　　　　　　　　　　 兰迪斯公司

Exercises

I. Read and judge.

(　　) 1. PLC is the only form of microcontroller.

(　　) 2. The term logic is used because programming is mainly concerned with implementing logic and switching operations.

(　　) 3. The early programmable logic controller was designed and developed by Modicon Company as a relay replacer for GM and Landis.

(　　) 4. The processor unit interprets the input signals and carries out control actions according to the program stored in its memory.

(　　) 5. The power supply unit is needed to convert the low DC voltage (5 V) to the mains AC voltage.

II. Translate the following phrases.

English	Chinese
1. programmable memory	
2. power supply unit	
3. General Motors	
4. output section	
5.	可编程逻辑控制器
6.	输入信号
7.	输出接口
8.	继电器

III. Fill in the blanks according to the text.

1. A _____ (PLC) is a special form of microcontroller. It uses _____ to store instructions and implement functions such as _____, _____, _____, _____, and _____ in order to control machines and processes.

2. Typically PLC has the basic components of processor unit, _____, _____, _____, communications and programming device.

3. The processor unit or central processing unit (CPU) is the unit containing the _____. This unit interprets the _____ and carries out _____ according to the program stored in its memory, communicating the decisions as _____ to the outputs.

4. The power supply unit is needed to convert the mains _____ to the low _____ (5 V) necessary for the processor and the circuits in the _____ interface modules.

IV. Complete the chart.

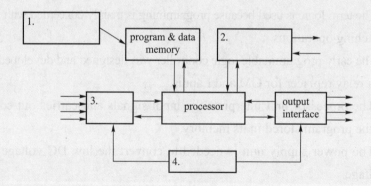

V. Translate the following sentences into Chinese.

1. PLC uses programmable memory to store instructions and implement functions such as logic, sequencing, timing, counting, and arithmetic in order to control machines and processes.

_____

2. The first PLC, model 084, was invented by Dick Morley in 1969.

_____

3. The processor unit is the unit containing the microprocessor.

_____

4. The programming device is used to enter the required program into the memory of the processor.

_____

Lesson 2 PLC operations

1. Basic operations of PLC

Read all input devices through the input interfaces, execute the user program stored in application memory, and then, based on the control scheme programmed by the user, turn the output devices on or off, or perform necessary control for the process application. This process of reading the inputs, executing the program in memory, and updating the outputs is known as scanning.

There are four basic steps in the operation (Fig.5.2.1) of all PLCs that are repeated many times per second: housekeeping, input scan, program scan and output scan. These steps continually take place in a repeating cycle.

Fig.5.2.1 Basic operation steps of PLC

1.1 Housekeeping

This step includes communications with programming terminals, internal diagnostics, etc.

1.2 Input scan

Input scan is to detect the state of all input devices that are connected to the PLC. When turned on the first time, it will check its own hardware and software for faults. If there are no problem, it will copy all the input and copy their values into memory, which is called the input scan.

1.3 Program scan

Program scan is to execute program logic created by the users. Using only the memory copy of the inputs, the ladder logic program will be solved once.

1.4 Output scan

When the program scan is done, the outputs will be updated using the temporary values in the memory, which is called the output scan.

2. Programming languages to program PLC

The PLC programming language international standard IEC 61131-3 deals with programming languages and defines two main kinds of programming languages including graphical programming language and text programming language.

2.1 Graphical programming language

Graphical programming language includes ladder diagram (LD), function block diagram (FBD) and sequential function chart (SFC).

- ***Ladder diagram (LD)***

Ladder logic/ladder diagram (LD) (Fig.5.2.2) is the most commonly used PLC programming language. It is a programming language similar to the relay circuit.

- ***Function block diagram (FBD)***

Function block diagram (FBD) (Fig.5.2.3) is a graphical language to describe signal and data flows through reusable function blocks. FBD is very useful for expressing the interconnection of control system algorithms and logic.

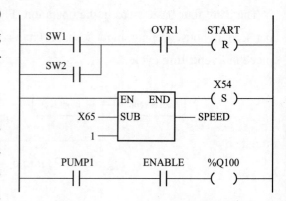

Fig.5.2.2 Ladder diagram (LD)

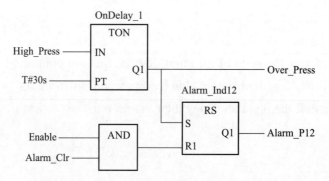

Fig.5.2.3 Function block diagram (FBD)

- ***Sequential function chart (SFC)***

Sequential function chart (SFC) (Fig.5.2.4) is a method of programming complex control systems at a more highly structured level. A SFC program is an overview of the control system.

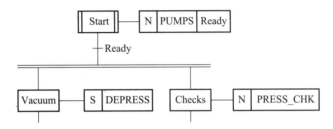

Fig.5.2.4   Sequential function chart (SFC)

2.2 Text programming language

Text programming language includes instruction list (IL) and structured text (ST).

- ***Instruction list (IL)***

Instruction list (IL) (Fig.5.2.5) is a low level programming language like assembly language. It consists of opcode and operand.

- ***Structured text (ST)***

Structured text (ST) (Fig.5.2.6) is a high level text language that encourages structured programming. It has a language structure that strongly resembles PASCAL or C programming language but simplified to some extent. It requires some knowledge on high level computer language and programming skills.

```
        LD    R1
        MPC   RESET
        LD    PRESS_1
        ST    MAX_PRESS
RESET:  LD    0
        ST    A_X43
```

```
If Speed1 > 100.0 then
    Flow_Rate: = 50.0 + Offset_A1;
Else
    Flow_Rate: = 100.0; Steam: = ON
End_If;
```

Fig.5.2.5   Instruction list (IL)   Fig.5.2.6   Structured text (ST)

New Words

5-2 单词

execute [ˈeksikjuːt]	*vt.*	执行；实施
scan [skæn]	*vt.*	扫描
housekeeping [ˈhausˌkiːpiŋ]	*n.*	内部处理
diagnostics [ˌdaiəgˈnɔstiks]	*n.*	诊断功能
fault [fɔːlt]	*n.*	缺点；故障
ladder [ˈlædə]	*n.*	梯子；阶梯
temporary [ˈtemprəri]	*adj.*	临时的；暂时的
graphical [ˈgræfikl]	*adj.*	图形的；绘画的
diagram [ˈdaiəgræm]	*n.*	图表；示意图

Unit 5 Programmable Logic Controller (PLC)

sequential [siˈkwenʃl]   adj. 顺序的；按次序的
interconnection [ˌintəkəˈnekʃn]   n. 相互连接
algorithm [ˈælgəriðəm]   n. 运算法则；计算程序
opcode [ˈɔpkəud]   n. 操作码
resemble [riˈzembəl]   vt. 与……相像；类似于

Phrases and Expressions

ladder logic   梯形逻辑
ladder diagram   梯形图
function block diagram   功能模块图
sequential function chart   顺序功能流程图
instruction list   指令表编程
structured text   结构化文本
PASCAL   结构化编程语言
C programming language   C 语言

Exercises

I. Read and judge.

(   ) 1. There are four basic steps in the operation of all PLCs that are repeated many times per second.

(   ) 2. Output scan includes communications with programming terminals, internal diagnostics, etc.

(   ) 3. Function block diagram is the most commonly used PLC programming language.

(   ) 4. Instruction list (IL) is a low level programming language like assembly language.

(   ) 5. Structured text is a low level text language.

II. Translate the following phrases.

English	Chinese
1. ladder diagram	
2. function block diagram	
3. sequential function chart	
4. instruction list	
5.	输入扫描
6.	程序扫描
7.	输出扫描
8.	文本编程语言

III. Fill in the blanks according to the text.

1. There are four basic steps in the operation of all PLCs that are repeated many times per second: _____, _____, _____ and _____.

2. Input scan is to detect the state of all input devices that are connected to the PLC. When turned on the first time, it will check its own hardware and software for _____.

3. Program scan is to _____ program logic created by the users.

4. The PLC programming language international standard IEC 61131-3 deals with programming languages and defines two main kinds of programming languages including _____ programming language and _____ programming language.

5. Graphical programming language includes _____ (LD), _____ (FBD) and _____ (SFC).

IV. Complete the chart according to the text.

```
[ 1. ] → [ 2. ] → [ 3. ] → [ 4. ] →
          ↑_____|
```

V. Translate the following sentences into Chinese.

1. This process of reading the inputs, executing the program in memory, and updating the outputs is known as scanning.
_____

2. Input scan is to detect the state of all input devices that are connected to the PLC.
_____

3. Program scan is to execute program logic created by the users.
_____

4. When the program scan is done, the outputs will be updated using the temporary values in the memory, which is called the output scan.
_____

5. Text programming language includes instruction list (IL) and structured text (ST).
_____

Lesson 3 Troubleshooting of PLC

1. Program troubleshooting

There are several causes of alteration to the user program:
(1) Extreme environmental conditions;
(2) Electromagnetic interference (EMI);
(3) Improper grounding;
(4) Improper wiring connections and unauthorized tampering.

If you suspect the memory has been altered, check the program against a previously saved program on an EEPROM or flash EPROM module.

2. Hardware troubleshooting

2.1 Troubleshooting of controller

In identifying the source of the controllers operation problem, we can use troubleshooting considerations table, including status indication, trouble description, probable causes and recommended action.

Follow these steps:
(1) Identify power supply and CPU LED status indicators (Fig.5.3.1);
(2) Match processor LEDs with the status LEDs located in troubleshooting tables;
(3) Once the status LEDs are matched to the appropriate table, simply move across the table identifying error description and probable causes.

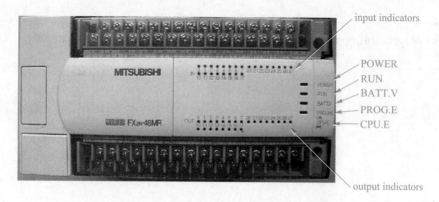

Fig.5.3.1 Diagnostic LED indicators

2.2 Power distribution

The master control relay must be able to stop all machines motion by removing power to the I/O devices when the relay is de-energized. The DC power supply should be powered directly from the fused secondary of the transformer. Power to the DC input and output, circuits are connected through a set of master control relay contacts. Interrupt the load side rather than the AC line power. This avoids the additional delay of power supply turn-on and turn-off.

2.3 Power LED

The power LED on the power supply indicates that DC power is being supplied to the chassis. This LED could be off when incoming power is present: ① when the fuse is blown; ② when voltage drops below the normal operating range; ③ when power supply is defective.

New Words

5-3 单词

troubleshooting [ˈtrʌblʃuːtiŋ]	n. 发现并修理故障;故障检测
alteration [ˌɔːltəˈreiʃn]	n. 变化;改变
unauthorized [ʌnˈɔːθəraizd]	adj. 未经授权的;未经许可的
tamper [ˈtæmpə]	vt. 篡改
alter [ˈɔːltə]	vt. 改变;变更
identify [aiˈdentifai]	vt. 识别,认出;确定
indication [ˌindiˈkeiʃən]	n. 指示
indicator [ˈindikeitə]	n. 指示器;指针
distribution [ˌdistriˈbjuːʃn]	n. 分配;分布
de-energize [dəˈenədʒaiz]	vt. 断开(电源)
fused [fjuːzd]	adj. 装有熔体的
delay [diˈlei]	n. 延迟
chassis [ˈʃæsi]	n. 底盘;底架
blow [bləu]	v. 使(熔体)熔断
normal [ˈnɔːml]	adj. 正常的
defective [diˈfektiv]	adj. 有错误的;有缺陷的

Phrases and Expressions

electromagnetic interference (EMI)	电磁干扰
troubleshooting considerations table	故障排除参考表
load side	载荷侧

Exercises

I. Read and judge.

(   ) 1. Extreme environmental conditions can't cause the alteration to the user program.

(   ) 2. If you suspect the memory has been altered, check the program against a previously saved program on an EEPROM or flash EPROM module.

(   ) 3. In identifying the source of the controllers operation problem, we can first match processor LEDs with the status LEDs located in troubleshooting tables.

(   ) 4. The master control relay must be able to stop all machines motion by removing power to the I/O devices when the relay is de-energized.

(   ) 5. The power LED on the power supply indicates that DC power is being supplied to the chassis.

II. Translate the following phrases.

English	Chinese
1. electromagnetic interference (EMI)	
2. power distribution	
3. secondary of the transformer	
4. load side	
5.	硬件故障检测
6.	故障排除参考表
7.	主控继电器
8.	额外延迟

III. Fill in the blanks according to the text.

1. We can use troubleshooting considerations table, including _____, _____, _____ and recommended action.

2. The _____ must be able to stop all machines motion by removing power to the I/O devices when the relay is _____.

3. The DC power supply should be powered directly from the fused _____.

IV. Translate the following sentences into Chinese.

1. In identifying the source of the controllers operation problem, we can use troubleshooting considerations table.

2. The master control relay must be able to stop all machines motion by removing power to the I/O devices when the relay is de-energized.

3. Interrupt the load side rather than the AC line power. This avoids the additional delay of power supply turn-on and turn-off.

4. The power LED on the power supply indicates that DC power is being supplied to the chassis.

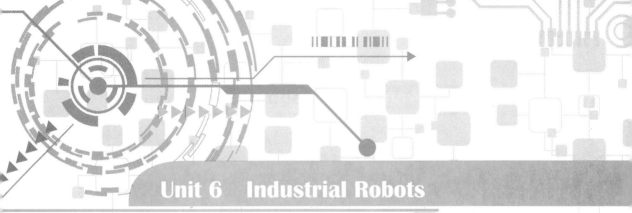

Unit 6 Industrial Robots

RS20   RS20 RS15L ReusL   Re13   Reus   Reus   Reus

In this unit, you will learn

◇ 1. Introduction to robots;
◇ 2. Types of industrial robots;
◇ 3. Typical applications of industrial robots.

Lesson 1　Introduction to robots

1. What is a robot

A robot is a multifunctional reprogrammable machine. A robot has been used to replace humans on the production line.

The superintelligent robots are able to reproduce the movements of human legs and arms. With a built-in brain, it will be able to make decisions and assemble parts. These machines will not only copy the movements of an operator in an exact manner, but will, with their built-in brain, have the ability to work without an operator.

The intelligent robot (Fig.6.1.1) possesses machine vision, a computer and sensors, the movements take place according to a defined program, as well as from instructions arriving from the sensors and the television camera.

There are also some non-intelligent robots as follows.

- Universal robot (Fig.6.1.2): This robot possesses an electric or electronic control. The arms move in three axes: vertical, horizontal and rotation. The movement can be from point-to-point or on a continuous path and can be carried out at the same time.

Fig.6.1.1　Intelligent robot

Fig.6.1.2　Universal robot

- Simple robot: This robot is simpler in design than the universal robot. It is designed for applications in certain areas, but is flexible without being universal.
- Miniature robot: These devices are used for assembling small parts.

2. Advantages and disadvantages of robots

2.1 Advantages of robots

- Robots work continuously without tiring or boredom.
- Robots can work in dangerous environments such as areas of radiation, darkness, too hot or cold, ocean bottoms (Fig.6.1.3), space (Fig.6.1.4) and so on without the need for life support, comfort, or concern for safety.

Fig.6.1.3 Robot underwater

Fig.6.1.4 Robot in space

- Robots have repeatable precision at all times unless something happens to them or unless they wear out.
- Robots and their accessories and sensors have capabilities beyond what humans could do.
- Robots can process multiple tasks at the same time.

2.2 Disadvantages of robots

- Robots lack capability to respond in emergencies, unless the situation is predicted or the response is included in the system.
- Robots have limited capabilities in cognition, creativity, decision-making and understanding.
- Robots have a high initial cost of equipment and installation. Sometimes, extra peripherals and programming are necessary.

New Words

6-1 单词

superintelligent [sju:pərinˈtelidʒənt]   *adj.* 超智能的
universal [ˌju:niˈvə:sl]   *adj.* 普遍的；通用的

vertical ['və:tikl] *adj.* 垂直的；竖立的
horizontal [ˌhɔri'zɔntl] *adj.* 水平的；卧式的
miniature ['minətʃə] *adj.* 小型的；微小的
boredom ['bɔ:dəm] *n.* 厌倦
environment [in'vairənmənt] *n.* 环境
radiation [ˌreidi'eiʃn] *n.* 辐射
repeatable [ri'pi:təbl] *adj.* 可重复的
precision [pri'siʒn] *n.* 精确度
accessory [ək'sesəri] *n.* 附件
multiple ['mʌltipl] *adj.* 多个的；多功能的
emergency [i'mə:dʒənsi] *n.* 紧急情况；突发事件
predict [pri'dikt] *vt.* 预测
cognition [kɔg'niʃn] *n.* 认知

Exercises

I. Read and judge.

(　　) 1. The superintelligent robots are able to reproduce the movements of human legs and arms.

(　　) 2. The superintelligent robots can not work without an operator.

(　　) 3. The universal robot possesses an electric or electronic control.

(　　) 4. The simple robot is not as flexible as the universal robot.

(　　) 5. Robots can process multiple tasks at the same time.

II. Translate the following phrases.

English	Chinese
1.	智能机器人
2.	通用机器人
3.	微型机器人
4. repeatable precision	
5. multiple tasks	
6. superintelligent robot	

III. Fill in the blanks according to the text.

1. A robot is a multifunctional _____ machine.
2. With a _____, the superintelligent robot will be able to make decisions.
3. The arms move in three axes: _____, _____ and _____.
4. Robots can work in _____ such as areas of radiation, darkness, too hot or cold.
5. Robots have a high initial cost of _____.

IV. Translate the following sentences into Chinese.

1. With their built-in brain, the superintelligent robots have the ability to work without an operator.

_____

2. Robots work continuously without tiring or boredom.

_____

3. Robots and their accessories and sensors have capabilities beyond what humans could do.

_____

4. Robots lack capability to respond in emergencies, unless the situation is predicted or the response is included in the system.

_____

5. Robots have limited capabilities in cognition, creativity, decision-making and understanding.

_____

Lesson 2 Types of industrial robots

6-2 课文

Industrial robots are automated, programmable and capable of movement on two or more axes. The major types of industrial robots by mechanical structure are:

1. Cartesian robot

With three prismatic joints, perpendicular to each other, cartesian robots (Fig.6.2.1) are easy to control and work with speeds above the average of other types of industrial robots.

The cartesian robots perform the linear movements according to a cartesian coordinate system *XYZ*. They are useful in picking and placing work, assembly operation, handling machine tools and arc welding.

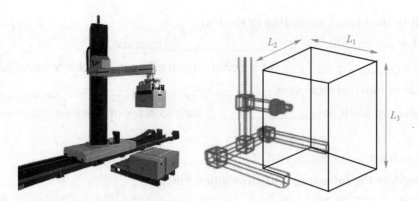

Fig.6.2.1 Cartesian robot and workspace

2. Cylindrical robot

The cylindrical robot (Fig.6.2.2) performs the movements and rotation according to a cylindrical coordinate system. Cylindrical robot is also used for assembly operation, handling machine tools, spot welding and handing die-casting machines.

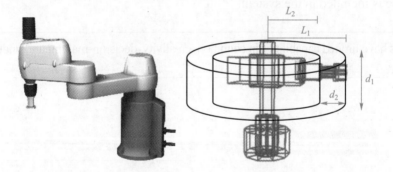

Fig.6.2.2 Cylindrical robot and workspace

3. Spherical robot

The spherical robot (Fig.6.2.3) is a robot with two rotary joints and one prismatic joint, in other words, two rotary axes and one linear axis. Spherical robots have an arm which forms a spherical coordinate.

This type of robot is used for handling machine tools, spot welding, die-casting, gas welding and arc welding.

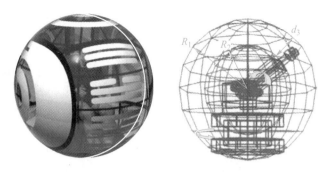

Fig.6.2.3 Spherical robot and workspace

4. SCARA robot

SCARA is short for selective compliant assembly robot arm or selective compliant articulated robot arm. The word "selective" refers to the flexibility of the robot in the *XY* plane and its rigidity in the third joint. It makes it unsuitable for working in planes involving the *Z* direction.

The SCARA robot (Fig.6.2.4) is used for picking and placing work, application of sealant, assembly operations and handling machine tools.

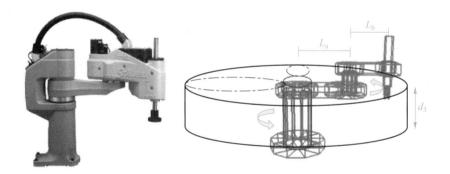

Fig.6.2.4 SCARA robot and workspace

5. Humanoid robot

In general, humanoid robots (Fig.6.2.5) have a body with a head, two arms and two legs, although some forms of humanoid robots may model only part of the body, for example, the hip, shoulder and human arm. The humanoid robot is the most versatile industrial robot.

With three revolute joints, the first axis perpendicular to the other two, the end effector of the robot can reach all the internal points of the length of arm. The humanoid robot is useful in many applications, especially those requiring careful operations around obstacles and access in narrow cavities.

Unit 6　Industrial Robots

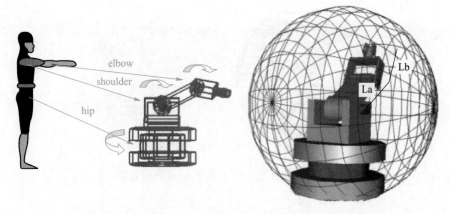

Fig.6.2.5　Joints and workspace of humanoid robot

6-2 单词

New Words

automated [ˈɔːtəmeitid]	*adj.* 自动化的
cartesian [kɑːˈtiːziən]	*adj.* 笛卡儿的
prismatic [prizˈmætik]	*adj.* 柱状的；棱柱的
joint [dʒɔint]	*n.* 关节；接头
perpendicular [ˌpəːpənˈdikjələ]	*adj.* 垂直的；成直角的
linear [ˈliniə]	*adj.* 直线的；线形的
coordinate [kəuˈɔːdineit]	*n.* 坐标
spherical [ˈsferikl]	*adj.* 球形的；球面的
selective [siˈlektiv]	*adj.* 选择的；有选择性的
compliant [kəmˈplaiənt]	*adj.* 遵从的；一致的
articulated [ɑːˈtikjuleitid]	*adj.* 铰接的；有关节的
unsuitable [ʌnˈsuːtəbl]	*adj.* 不适合的
involve [inˈvɔlv]	*vt.* 包含；涉及
sealant [ˈsiːlənt]	*n.* 密封剂
humanoid [ˈhjuːmənɔid]	*adj.* 仿人的；有人的特点的
hip [hip]	*n.* 臀部
shoulder [ˈʃəuldə]	*n.* 肩膀
versatile [ˈvəːsətail]	*adj.* 多用途的
effector [iˈfektə]	*n.* 效应器；执行器
obstacle [ˈɔbstəkl]	*n.* 障碍物
cavity [ˈkævəti]	*n.* 腔；型腔

Phrases and Expressions

cartesian robot	笛卡儿机器人
picking and placing	拾取
cartesian coordinate system	笛卡儿坐标系
cylindrical robot	圆柱坐标机器人
die-casting machine	压铸机
spherical robot	球形机器人
selective compliant assembly robot arm	选择顺应性装配机器手臂
the humanoid robot	仿人机器人

Exercises

I. Read and judge.

(   ) 1. With three prismatic joints, perpendicular to each other, cartesian robots are easy to control and work with speeds.

(   ) 2. Spherical robots have an arm which forms a spherical coordinate.

(   ) 3. The SCARA robot is suitable for working in planes involving the X, Y and Z direction.

(   ) 4. The SCARA robot is the most versatile industrial robot.

(   ) 5. In general, humanoid robots have a body with a head, two arms and two legs.

II. Translate the following phrases.

English	Chinese
1.	工业机器人
2.	笛卡儿机器人
3.	圆柱坐标机器人
4.	压铸机
5. cartesian coordinate system	
6. selective compliant assembly robot arm	
7. spherical robot	
8. humanoid robot	

III. Fill in the blanks according to the text.

1. With three _____, perpendicular to each other, cartesian robots are easy to

control and work with speeds above the average of other types of industrial robots.

2. The cartesian robots are useful in _____ work, assembly operation, handling machine tools and _____.

3. Cylindrical robot is also used for assembly operation, handling machine tools, _____ and handing _____.

4. SCARA is short for _____ or selective compliant articulated robot arm. The word "selective" refers to the _____ of the robot in the *XY* plane and its _____ in the third joint.

5. The _____ is useful in many applications, especially those requiring careful operations around obstacles and access in narrow cavities.

IV. Translate the following sentences into Chinese.

1. The cartesian robots perform the linear movements according to a cartesian coordinate system *XYZ*.
_____

2. The cylindrical robot performs the movements and rotation according to a cylindrical coordinate system.
_____

3. This type of robot is used for handling machine tools, spot welding, die-casting, gas welding and arc welding.
_____

4. The humanoid robot is the most versatile industrial robot.
_____

5. The end effector of the robot can reach all the internal points of the length of arm.
_____

Lesson 3 Typical applications of industrial robots

Typical applications of industrial robots include welding, painting, assembly, picking and placing, packaging and labeling, palletizing, product inspection and testing; all accomplished with high endurance, speed and precision.

1. Robot welding

Robot welding is to completely automate a welding process by the use of the robots. Processes such as gas metal arc welding, while often automated, are not necessarily equal to robot welding, since a human operator sometimes prepares the welding materials. Robot welding is commonly used for resistance spot welding (Fig.6.3.1) and arc welding (Fig.6.3.2) in high production applications, such as the automotive industry.

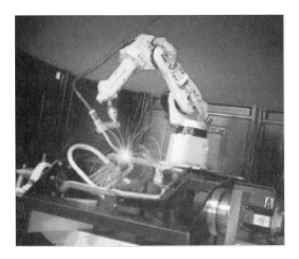

Fig.6.3.1 Robot spot welding   Fig.6.3.2 Robot arc welding

The major components of arc welding robots are the manipulator or the mechanical unit and the controller, which acts as the robot's "brain". The manipulator is what makes the robot move and the design of these systems can be grouped into several common types, such as SCARA and cartesian coordinate robot, which use different coordinate systems to direct the arms of the machine.

2. The spray painting robot

A spray painting robot (Fig.6.3.3) is also called painting robot. This kind of industrial robot is used to apply spray painting or other coating automatically. The spray painting robot consists of the robot body, a computer and other control systems.

The painting robot is generally driven by hydraulic pressure, which has the advantages of high speed and good explosion-proof performance. The spraying robot can work continuously in the harsh environment, working flexibly and high-accurately. Therefore, the

Fig.6.3.3 Robot spray painting

painting robot is widely used in painting production lines such as automobiles, large structural parts and so on.

3. The assembly robot

The assembly robot (Fig.6.3.4) is the core equipment of flexible automatic assembly system, which is composed of robot operator, controller, terminal actuator and sensor system. The assembly robot has the characteristics of high precision, good flexibility and small working range, and can be used with other systems. It is mainly used in various electrical appliance manufacturing industries.

4. The palletizing robot

There are only two points in a palletizing robot's computer to be located. One is the starting point and the other is the placement point. The computer will search for the most suitable orbit for these two points to move. The palletizing robots (Fig.6.3.5) are widely used in chemical, beverage, food, beer, plastic and other automatic manufacturing industries. They are suitable for cases, bags, cans, bottles and other forms of packaging.

Fig.6.3.4　Assembly robot

Fig.6.3.5　Palletizing robot

6-3 单词

New Words

package ['pækidʒ]　　　　　　　　　vt. 包装
palletize ['pælitaiz]　　　　　　　　vt. 码垛堆集

endurance [inˈdjuərəns]   n. 耐久力；持久力
manipulator [məˈnipjuleitə]   n. 机械臂；操纵器
coating [ˈkəutiŋ]   n. 涂层
explosion [ikˈspləuʒn]   n. 爆炸
proof [pru:f]   adj. 防……的；抗……的
harsh [hɑ:ʃ]   adj. 残酷的
core [kɔ:]   n. 核心
characteristic [ˌkærəktəˈristik]   n. 特性；特征
actuator [ˈæktʃueitə]   n. 执行机构；促动器
locate [ləuˈkeit]   vi. 定位
orbit [ˈɔ:bit]   n. 轨道
beverage [ˈbevəridʒ]   n. 饮料
can [kæn]   n. 罐头

Phrases and Expressions

robot welding   机器人焊接
spray painting robot   喷涂机器人
hydraulic pressure   液压
explosion-proof performance   防爆性能
painting production line   涂装生产线
flexible automatic assembly system   柔性自动装配系统
terminal actuator   终端执行器
palletizing robot   码垛机器人

Exercises

I. Read and judge.

(　　) 1. Gas metal arc welding is a kind of robot welding.

(　　) 2. A spray painting robot is used to apply spray painting or other coating automatically.

(　　) 3. The spraying robot can work continuously in the harsh environment.

(　　) 4. The assembly robot should work in a large working range.

(　　) 5. There are only two points in a palletizing robot's computer to be located.

II. Translate the following phrases.

English	Chinese
1.	机器人焊接
2.	喷涂机器人
3.	涂装生产线
4.	码垛机器人
5. hydraulic pressure	
6. explosion-proof performance	
7. flexible automatic assembly system	
8. terminal actuator	

III. Fill in the blanks according to the text.

1. Robot welding is commonly used for resistance _____ and _____ in high production applications, such as the _____.

2. The major components of arc welding robots are the _____ or the mechanical unit and the _____, which acts as the robot's "brain".

3. A spray painting robot is also called _____.

4. The assembly robot is composed of robot operator, controller, _____ and _____.

5. The palletizing robots are widely used in chemical, _____, food, beer, plastic and other _____ industries.

IV. Translate the following sentences into Chinese.

1. Robot welding is to completely automate a welding process by the use of the robots.

2. The painting robot is generally driven by hydraulic pressure, which has the advantages of high speed and good explosion-proof performance.

3. The assembly robot is the core equipment of flexible automatic assembly system.

4. The assembly robot has the characteristics of high precision, good flexibility and small working range, and can be used with other systems.

5. The palletizing robots are suitable for cases, bags, cans, bottles and other forms of packaging.

Unit 7   General View of CNC

In this unit, you will learn

◇ 1. Introduction to CNC;
◇ 2. Applications of CNC;
◇ 3. Safety requirements of CNC operations.

Lesson 1　Introduction to CNC

1. What is CNC

CNC (computer numerical control) is the process of manufacturing machine parts. The production is controlled by a computerized controller. The controller uses motors to drive each axis of a machine tool and controls the direction, speed and length of time when each axis rotates. A programmed path is loaded into the machine's computer by the operator and then executed.

2. History of CNC

To ensure that all United States military airplanes were manufactured identically, the United States Air Force invited several companies to develop and manufacture a numerical system that could handle the volume and repeatability of machine parts.

The first contract was awarded to the Parsons Corporation of Michigan. In 1951, the Servomechanism Laboratory of MIT (the Massachusetts Institute of Technology) (Fig.7.1.1) was given a subcontract by Parson to develop a servo system for machine tools. In 1952, the first three-axis, numerically controlled machine tool (Fig.7.1.2) was created.

Fig.7.1.1　Servomechanism Laboratory of MIT

CNC technology has been one of the manufacturing's major developments in the past 50 years. It has not only resulted in the development of new techniques and the achievement of higher production levels, but also helped increase product quality and reduce manufacturing costs.

Fig.7.1.2　The first CNC machine producing identical parts

3. The workflow in CNC

There is a common workflow for designing and manufacturing a part by CNC:
- A part is designed in a CAD (computer aided design) program.
- The design is loaded into a CAM (computer aided manufacturing) program and translated into instructions called G-code.
- The G-code is checked for correctness and fixed if necessary.
- The G-code instructions are loaded into a CNC controller program which sends signals to electronic circuit which controls the motors on the CNC machine.
- The CNC machine cuts the parts from a block of material.

New Words

7-1 单词

numerical [njuːˈmerɪkl]	*adj.* 数字的；数值的
computerized [kəmˈpjuːtəˌraɪzd]	*adj.* 计算机化的
axis [ˈæksɪs]	*n.* 轴；轴线
military [ˈmɪlətri]	*adj.* 军用的；军事的
volume [ˈvɒljuːm]	*n.* 量；大量
repeatability [rɪˌpiːtəˈbɪlɪti]	*n.* 可重复性；反复性
contract [ˈkɒntrækt]	*n.* 合同；契约
award [əˈwɔːd]	*vt.* 授予；奖给

servomechanism [ˈsəːvəuˌmikəˌnizəm]　　　　*n.* 伺服机构系统

Phrases and Expressions

computer numerical control (CNC)	计算机数字控制
machine tool	机床
United States Air Force	美国空军
Massachusettes Institute of Technology (MIT)	麻省理工学院
servo system	伺服系统

Exercises

I. Read and judge.

(　　) 1. The production is controlled by a computerized controller.

(　　) 2. The controller uses a computer to drive each axis of a machine tool.

(　　) 3. The first three-axis, numerically controlled machine tool was created to help industry.

(　　) 4. CNC technology has been one of the manufacturing's major developments in the past 50 years.

(　　) 5. The G-code is checked for correctness and fixed if necessary.

II. Translate the following phrases.

English	Chinese
1.	计算机数字控制
2.	机床
3.	伺服机构系统
4. length of time	
5. a programmed path	
6. product quality	

III. Fill in the blanks according to the text.

1. The production is controlled by a computerized _____. The controller uses motors to drive each _____ of a machine tool and controls the direction, speed and _____ when each axis rotates. _____ is loaded into the machine's computer by the operator and then executed.

2. The design is loaded into a CAM program and translated into instructions called _____.

3. The G-code _____ are loaded into a CNC controller program.

IV. Translate the following sentences into Chinese.

1. CNC (computer numerical control) is the process of manufacturing machine parts.
_____

2. The controller uses motors to drive each axis of a machine tool and controls the direction, speed and length of time when each axis rotates.
_____

3. A part is designed in a CAD (computer aided design) program.
_____

4. The design is loaded into a CAM (computer aided manufacturing) program and translated into instructions called G-code.
_____

5. The G-code instructions are loaded into a CNC controller program.
_____

Lesson 2 Applications of CNC

7-2 课文

1. Types of CNC machine tools

There are many different types of CNC machines. The number of installations is rapidly increasing, and the technology development advances at a rapid pace. Here is a brief list of the groups of CNC machines:
- Lathe and turning center;
- Mill and machining center;
- Drilling machine;
- Boring machine;
- Wire cut electrical discharge machine (EDM);
- Flame cutting machine (Fig.7.2.1);
- Cylindrical grinder;
- Welding machine, etc.

Fig.7.2.1 Flame cutting machine

2. Applications of CNC machine tools

2.1 Industries for removing metal

The metal removing industries remove metal from raw materials to give the materials desired shapes as requirements. These can be the automotive industries for making the shafts, gears, and many other parts. It can also be used by manufacturing industries for making the various rounded, square, threaded parts and other jobs.

2.2 Industries for fabricating metals

In many industries, thin plates like steel plates are required for various purposes. In fabricating industries, the CNC machines are used for various machining operations like shearing, flame cutting, punching, forming and welding on the thin plates.

2.3 Electrical discharge machining (EDM) industry

Electrical discharge machining (EDM) is a method of removing metal with the use of electrical sparks. These electrical discharge machines have two types, the sinker EDM and wire cut EDM. The sinker EDM (Fig.7.2.2) consists of an electrode and workpieces submerged in an insulating liquid such as oil. Wire cut EDM (Fig.7.2.3) uses a thin single-strand metal wire (usually brass) to cut plates as thick as 300 mm.

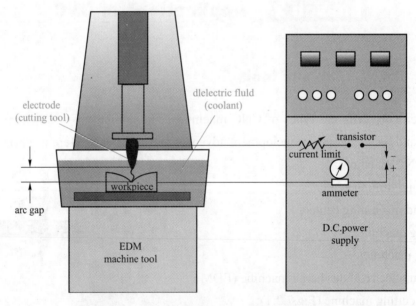

Fig.7.2.2 Sinker EDM

2.4 Other industries where CNC machines are used

CNC machines are also used widely in the wood working industries to perform various operations. CNC technology is also used in number of lettering and engraving systems. There are also CNC machines for the electrical industries such as CNC coil winders (Fig.7.2.4) and soldering machines.

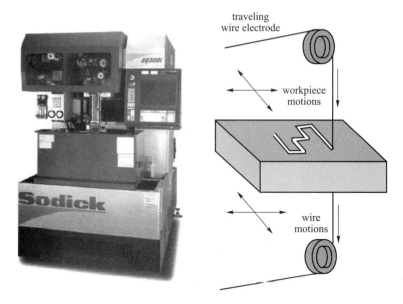

Fig.7.2.3 Wire cut EDM

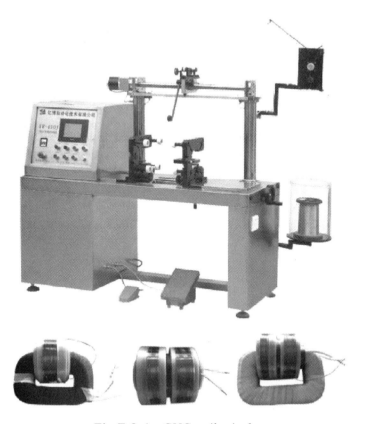

Fig.7.2.4 CNC coil winder

7-2 单词

New Words

lathe [leið]	n. 车床
mill [mil]	n. 铣床；磨坊
drilling [ˈdriliŋ]	n. 钻孔
boring [ˈbɔːriŋ]	n. 镗孔
flame [fleim]	n. 火焰
grinder [ˈgraində]	n. 磨床
raw [rɔː]	adj. 生的；未加工的
fabricate [ˈfæbrikeit]	vt. 制造
shear [ʃiə]	vt. 剪切；切断
punch [pʌntʃ]	vt. 打孔；穿孔，冲孔
spark [spɑːk]	n. 火花
electrode [iˈlektrəud]	n. 电极；电焊条
submerge [səbˈməːdʒ]	vt. & vi. 淹没；把……浸入
liquid [ˈlikwid]	n. 液体
lettering [ˈletəriŋ]	n. 刻字
engrave [inˈgreiv]	vt. 雕刻；镌刻
coil [kɔil]	n. 线圈

Phrases and Expressions

turning center	车削加工中心
machining center	加工中心
drilling machine	钻床
boring machine	镗床
wire cut electrical discharge machine	线切割电火花机
flame cutting machine	火焰切割机
cylindrical grinder	外圆磨床
insulating liquid	绝缘液
engraving systems	雕刻系统
coil winder	线圈缠绕机
soldering machine	钎焊机

Exercises

I. Read and judge.

(   ) 1. In many industries, thin plates like steel plates are required for various purposes.

(   ) 2. These electrical discharge machines have two types, the sinker EDM and wire cut EDM.

(   ) 3. The sinker EDM consists of an electrode and workpieces submerged in an insulating liquid such as water.

(   ) 4. Wire cut EDM uses a thin single-strand metal wire to cut thin plates.

(   ) 5. CNC machines are also used widely in the wood working industries.

II. Translate the following phrases.

English	Chinese
1.	车削加工中心
2.	加工中心
3.	火焰切割机
4.	钎焊机
5. wire cut electrical discharge machining	
6. cylindrical grinder	
7. insulating liquid	
8. coil winder	

III. Fill in the blanks according to the text.

_____ (EDM) is a method of removing metal with the use of electrical sparks. These electrical discharge machines have two types, the _____ EDM and _____ EDM. The sinker EDM consists of an electrode and workpieces submerged in an _____ such as oil. Wire cut EDM uses a thin _____ (usually brass) to cut plates as thick as 300 mm.

IV. Translate the following sentences into Chinese.

1. The metal removing industries remove metal from raw materials to give the materials desired shapes as requirements.

_____

2. Electrical discharge machining (EDM) is a method of removing metal with the use of electrical sparks.

_____

3. These electrical discharge machines have two types, the sinker EDM and wire cut EDM.

4. The sinker EDM consists of an electrode and workpieces submerged in an insulating liquid such as oil.

5. There are also CNC machines for the electrical industries such as CNC coil winders and soldering machines.

Lesson 3 Safety requirements of CNC operations

In CNC operations, operators should know about the safety operation requirements in order to prevent or avoid the personnel injuries and the equipment damages (Fig.7.3.1).

Fig.7.3.1　Safety operation

1. Requirements of basic operations

- Don't touch transformers, electrical motors, junction boxes and some other processing terminal positions.
- Don't touch the switch with moist hands.
- Be familiar with the placement of the emergency stop button.
- Cut off the power, when there is any accident.
- Examine the measurement of lubricant (Fig.7.3.2), coolant and hydraulic fuel, and add

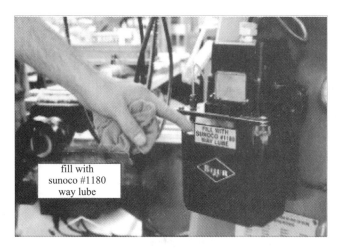

Fig.7.3.2 Examine the measurement of lubricant

them when needed.
- Don't alter the parameter, value and other electrical equipment casually. If it must be altered, record the original data in order to recover the original data when finished.

2. Requirements in work

- Wear the safety protection facilities, such as work clothes (Fig.7.3.3).
- Please tie back the long hair and wear the hat when operating (Fig.7.3.4).
- Clamp the workpiece.
- Adjust the jet nipple of the coolant under the stop condition.
- Don't touch the rotating workpiece and the principal axis by hands.
- Don't open the protective door under the automatic processing.

Fig.7.3.3 Wear work clothes

Fig.7.3.4 Tie back the long hair

3. Requirements after work

- Don't clean the machine before it stops working.
- After finishing, clean the cutting metal, door, lid, window, etc.
- Return each part to the original placement.
- Examine the pollution situation of the coolant, lubricant and hydraulic fuel. If the situation is severe, please change at once.

7-3 单词

New Words

prevent [priˈvent]	vt. 预防；阻止
damage [ˈdæmidʒ]	n. 损害
transformer [trænsˈfɔːmə]	n. 变压器
junction [ˈdʒʌŋkʃn]	n. 连接；接合
moist [mɔist]	adj. 潮湿的；湿润的
lubricant [ˈluːbrikənt]	n. 润滑剂；润滑油
coolant [ˈkuːlənt]	n. 冷却剂；防冻液
hydraulic [haiˈdrɔːlik]	adj. 液压的；水压的
fuel [ˈfjuːəl]	n. 燃料；燃油
parameter [pəˈræmitə]	n. 参数
recover [riˈkʌvə]	vt. 恢复

Phrases and Expressions

safety operation	安全操作
junction box	接线盒；分线箱
emergency stop button	急停按钮
jet nipple	喷嘴
principal axis	主轴
protective door	防护门

Exercises

I. Read and judge.

( ) 1. Don't touch the switch with moist hands.

( ) 2. Never alter the parameter, value and other electrical equipment.

( ) 3. Wear the safety protection facilities, such as work clothes.

( ) 4. Adjust the jet nipple of the coolant liquid as you like.

( ) 5. After finishing, clean the cutting metal, door, lid, window, etc.

II. Translate the following phrases.

English	Chinese
1.	安全操作
2.	接线盒
3.	急停按钮
4.	工作服
5. cut off the power	
6. jet nipple	
7. principal axis	
8. protective door	

III. Fill in the blanks according to the text.

1. In CNC operations, operators should know about the _____ requirements in order to prevent or avoid the personnel injuries and the equipment damages.

2. _____, when there is any accident.

3. Examine the measurement of lubricant, _____ and hydraulic fuel, and add them when needed.

4. Don't alter the _____, _____ and other electrical equipment casually. If it must be altered, record the original data in order to _____ the original data when finished.

5. Wear the safety protection facilities, such as _____.

6. Adjust the _____ of the coolant under the stop condition.

IV. Translate the following sentences into Chinese.

1. Don't touch transformers, electrical motors, junction boxes and some other processing terminal positions.

_____

2. Don't touch the switch with moist hands.

_____

3. Be familiar with the placement of the emergency stop button.

4. Don't open the protective door under the automatic processing.

5. Return each part to the original placement.

Unit 8 CNC Machine Tools

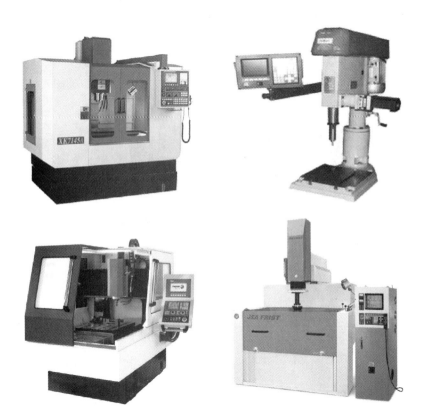

In this unit, you will learn

◇ 1. CNC lathe and turning operations;
◇ 2. Milling machines and milling operations;
◇ 3. Machining centers.

Lesson 1 CNC lathe and turning operations

A lathe is a machine tool that can remove materials from a workpiece (which is held by a spindle) by a cutter.

1. Components of CNC lathe

A CNC lathe generally consists of the following components (Fig.8.1.1):

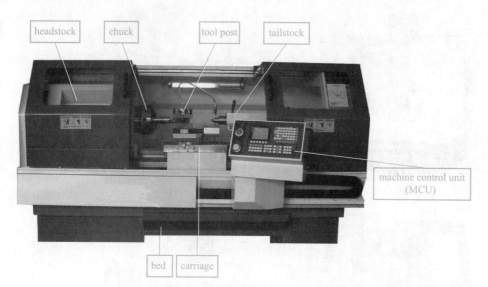

Fig.8.1.1 Components of a CNC lathe

1.1 Headstock

The headstock is the front section of the machine that contains the motor and drive system which powers the spindle. The spindle supports and rotates the workpiece which is clamped in a workpiece holder, such as a chuck.

1.2 Chuck

The chuck is connected with the spindle and the workpiece is clamped in it.

1.3 Tool post

The tool post is used to put the tools on the machine.

1.4 Tailstock

The tailstock is the rear section of the machine. It is used to support the other end of the workpiece and allow it to rotate. It can slide along the ways and be clamped at any position.

1.5 Machine control unit (MCU)

The machine control unit (MCU) is used to store and process CNC programs.

1.6 Carriage

The carriage moves the cutting tool to cut the rotating workpiece.

1.7 Bed

The bed supports all the structures listed above.

2. Basic lathe operations

Basic lathe operations are turning, facing, parting, grooving, drilling, boring and threading.

2.1 Turning

Turning (Fig.8.1.2) is to remove materials from the outer round surface of a rotating workpiece. A rough cut pass is usually made first. This is followed by one or more finishing passes.

2.2 Facing

Facing (Fig.8.1.3) is to cut the end (face) of the workpiece to make it flat.

Fig.8.1.2 Turning   Fig.8.1.3 Facing

2.3 Parting (cutting off)

Parting (Fig.8.1.4) is to cut off the part from the workpiece, such as a main bar.

2.4 Grooving

There are two types of grooves that can be made on the lathe. The first is similar to parting—a groove into the outer round surface of a workpiece, while the second is on the face of a workpiece (Fig.8.1.5).

Fig.8.1.4 Parting

Fig.8.1.5 Grooving

2.5 Drilling

Drilling (Fig.8.1.6) is to drill holes on the lathe. A drill is usually mounted in a drill chuck or held in a bushing and fed into the rotating workpiece.

2.6 Boring

Boring (Fig.8.1.7) is an internal turning operation.

Fig.8.1.6 Drilling

Fig.8.1.7 Boring

2.7 Threading

Threading (Fig.8.1.8) is used to cut helical grooves on the outside or inside surface of the workpiece.

Lesson 1 CNC lathe and turning operations

Fig.8.1.8   Threading

New Words

8-1 单词

remove [riˈmuːv]	*vt.* 移除；去除
spindle [ˈspindl]	*n.* 主轴
headstock [ˈhedstɔk]	*n.* 主轴箱
chuck [tʃʌk]	*n.* 卡盘
tailstock [ˈteilstɔk]	*n.* 尾座
carriage [ˈkæridʒ]	*n.* 滑板
turning [ˈtəːniŋ]	*n.* 车削；切削外圆
facing [ˈfeisiŋ]	*n.* 端面车削；车端面
parting [ˈpɑːtiŋ]	*n.* 切断
grooving [ˈgruːviŋ]	*n.* 切槽
threading [ˈθrediŋ]	*n.* 车螺纹；攻螺纹
surface [ˈsəːfis]	*n.* 表面
helical [ˈhelikl]	*adj.* 螺旋状的

Phrases and Expressions

drive system	驱动系统
workpiece holder	工件夹具
tool post	刀架
machine control unit	机床控制单元
rough cut pass	粗加工
finishing pass	精加工
cutting off	切断
main bar	主棒料
be similar to	与……相似
helical groove	螺旋槽

Unit 8 CNC Machine Tools

Exercises

I. Write down the names of the components of a CNC lathe.

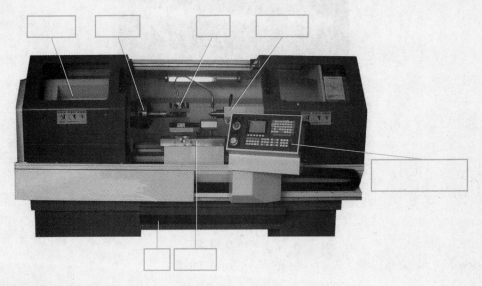

II. Translate the following phrases.

English	Chinese
1.	机床控制单元
2.	工件夹具
3.	切断
4. drive system	
5. main bar	
6. helical grooves	

III. Fill in the blanks according to the text.

1. A lathe is a machine tool that can remove materials from a _____ by a cutter.

2. The spindle supports and rotates the workpiece which is clamped in a _____, such as a chuck.

3. The _____ is the rear section of the machine. It is used to support the other end of the workpiece and allow it to rotate.

4. The _____ moves the cutting tool to cut the rotating workpiece.

5. _____ is to cut the end (face) of the workpiece to make it flat.

6. _____ is to cut off the part from the workpiece, such as a main bar.

7. There are two types of grooves that can be made on the lathe. The first is similar

to parting—a groove into the _____ of a workpiece, while the second is on the _____ of a workpiece.

8. _____ is an internal turning operation.

IV. Match the pictures with the words in the box.

| parting   threading   turning   facing   boring   grooving |

1. _____   2. _____   3. _____

4. _____   5. _____   6. _____

V. Translate the following sentences into Chinese.

1. The chuck is connected with the spindle and the workpiece is clamped in it.

2. The machine control unit (MCU) is used to store and process CNC programs.

3. Basic lathe operations are turning, facing, parting, grooving, drilling, boring and threading.

4. Turning is to remove materials from the outer round surface of a rotating workpiece.

5. Threading is used to cut helical grooves on the outside or inside surface of the workpiece.

Lesson 2 Milling machines and milling operations

1. What's a milling machine

A milling machine (Fig.8.2.1) is often called a mill. The term miller was commonly used in the 19th and early 20th centuries. Since the 1960s, the term milling machine has developed.

A milling machine is a machine tool which removes metal from the workpiece with a rotating milling cutter as the workpiece is fed against it. Milling machines are considered the most widely used machine tools.

Fig.8.2.1 Milling machine

2. Types of milling machines

Milling machines can be divided into three categories.

2.1 By the orientation of axes — vertical or horizontal

A mill where the movement of the spindle is up and down is a vertical milling machine (Fig.8.2.2). A mill where the movement of the spindle is in and out is a horizontal milling machine (Fig.8.2.3).

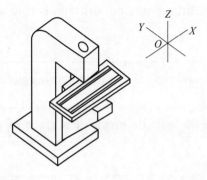
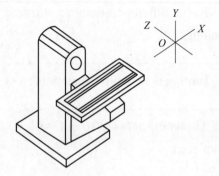

Fig.8.2.2 Vertical milling machine   Fig.8.2.3 Horizontal milling machine

2.2 By the number of axes — three, four, five or more

There are nine standard axes (Fig.8.2.4) used in CNC machining, including three basic axes: X, Y and Z to identify linear movement; three rotary axes: A, B and C to identify arc or circular movement; three parallel axes: U, V and W to identify auxiliary linear movement.

Usually, a lathe has two axes—X and Z. Milling machines and machining centers have at least three basic axes—X, Y and Z. The machines become even more flexible if they have a fourth axis, usually a rotary axis—A, B or C. Even higher level of flexibility can be found on machines with five or more axes (Fig.8.2.5).

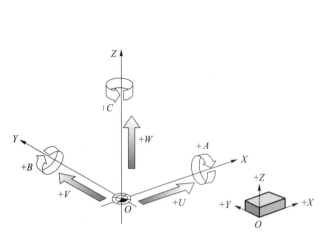

Fig.8.2.4 Nine standard axes

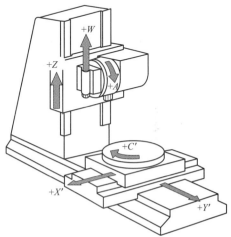

Fig.8.2.5 Five or more axes

2.3 With or without an automatic tool changer (ATC) — milling machine or machining center

There is no automatic tool changer (ATC) (Fig.8.2.6) in the milling machine. Mills with automatic tool changers are machining centers, which hold many tools. When required, the required tool can be automatically placed in the spindle for machining.

Fig.8.2.6 Automatic tool changer (ATC)

3. Milling operations

- Flat surface (Fig.8.2.7) and formed surface (Fig.8.2.8) may be machined excellently because table movements have micrometer adjustments.

Fig.8.2.7   Flat surface

Fig.8.2.8   Formed surface

- Angles, slots and gear teeth can be made with various cutters.
- Drills and boring tools can be held in the tool holders to do drilling and boring.
- Most operations performed on shapers, drilling machines, gear-cutting machines can be done on the milling machines.

8-2 单词

New Words

feed [fi:d]	*vt.* 进给；喂养
orientation [ˌɔːriənˈteiʃn]	*n.* 方向；定位
parallel [ˈpærəlel]	*adj.* 平行的
	n. 平行线
auxiliary [ɔːgˈziliəri]	*adj.* 辅助的
flexibility [ˌfleksəˈbiləti]	*n.* 灵活性
micrometer [maiˈkrɔmitə]	*n.* 微米
adjustment [əˈdʒʌstmənt]	*n.* 调节；调整
slot [slɔt]	*n.* 槽
shaper [ˈʃeipə]	*n.* 刨床

Phrases and Expressions

milling cutter	铣刀
vertical milling machine	立式铣床
horizontal milling machine	卧式铣床
linear movement	直线运动
arc or circular movement	圆弧或圆周运动
auxiliary linear movement	辅助直线运动

automatic tool changer (ATC)　　　自动换刀装置
flat surface　　　平面
formed surface　　　成形面
tool holder　　　刀具柄；刀具夹
gear-cutting machine　　　切齿机

Exercises

I. Read and judge.

(　　) 1. Milling machines are considered the most widely used machine tools.

(　　) 2. There are six standard axes used in CNC machining.

(　　) 3. A lathe has three axes—X, Y and Z.

(　　) 4. The machines become even more flexible if they have a fourth axis, usually a rotary axis.

(　　) 5. Mills with automatic tool changers are machining centers, which hold many tools.

II. Translate the following phrases.

English	Chinese
1.	铣刀
2.	立式铣床
3.	卧式铣床
4.	自动换刀装置
5. linear movement	
6. arc or circular movement	
7. auxiliary linear movement	
8. formed surface	

III. Fill in the blanks according to the text.

1. A milling machine is often called a _____. The term _____ was commonly used in the 19th and early 20th centuries. Since the 1960s, the term _____ has developed.

2. A milling machine is a machine tool which removes metal from the workpiece with a rotating _____ as the _____ is fed against it.

3. There are nine standard axes used in CNC machining, including three_____: X, Y and Z to identify linear movement; three _____: A, B and C to identify

arc or circular movement; three _____: U, V and W to identify auxiliary linear movement.

IV. Translate the following sentences into Chinese.

1. A milling machine is a machine tool which removes metal from the workpiece with a rotating milling cutter as the workpiece is fed against it.

_____

2. Three basic axes: X, Y and Z to identify linear movements.

_____

3. Flat surface and formed surface may be machined excellently because table movements have micrometer adjustments.

_____

4. Angles, slots and gear teeth can be made with various cutters.

_____

5. Most operations performed on shapers, drilling machines, gear-cutting machines can be done on the milling machines.

_____

Lesson 3 Machining centers

1. What is a machining center

CNC machining centers developed from milling machines. In 1968, an NC (numerical control) machine was marketed which could automatically change tools so that many different processes could be done on one machine.

A machining center (Fig.8.3.1) is a mill with an automatic tool changer (ATC) that includes a tool magazine, and sometimes an automatic pallet changer (APC).

Fig.8.3.1   Machining center

2. Tool magazine

All CNC machining centers have the automatic tool changers including a tool magazine

(Fig.8.3.2). A tool magazine consists of a certain number of pockets where the tool holders with cutting tools are placed. The capacity of a magazine can be as small as ten or twelve tools and as large as several hundred tools. The medium size machining center may have 20~40 tools.

Fig.8.3.2 Tool magazine

3. Four-axis machining center

A four-axis machining center has X-, Y- and Z-axis as basic axes, plus a rotary axis / table, usually A-axis or C-axis for vertical machining centers (Fig.8.3.3) and B-axis for horizontal machining centers (Fig.8.3.4).

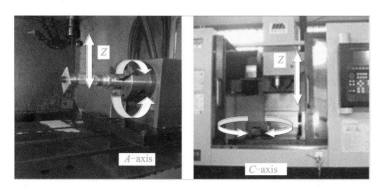
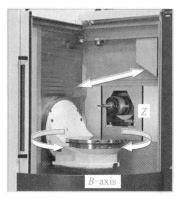

Fig.8.3.3 Four-axis vertical machining centers

Fig.8.3.4 Four-axis horizontal machining center

A machining center with four-axis control can automatically select and change as many as 32 preset tools.

4. Five-axis machining center

4.1 Five-axis machining center

The most advanced CNC machining centers are designed to add two more axes (*A*-, *B*- or *C*-axis) in addition to the three basic axes (*X*-, *Y*- and *Z*-axis) (Fig.8.3.5). When all of these axes are used in conjunction with each other, extremely complicated geometries can be made with ease.

Fig.8.3.5　Five-axis machining center

4.2 Advantages of five-axis machining

- Reduce machining time and cost.
- Better surface finish.
- No need to re-position the workpiece at complicated angles.
- Machine much more complicated parts, including normal holes required on a complicated surface.

8-3 单词

New Words

magazine [ˌmæɡəˈziːn]	*n.* 刀库；弹药库；杂志
pallet [ˈpælət]	*n.* 平台；拖盘
pocket [ˈpɔkit]	*n.* 口袋
capacity [kəˈpæsəti]	*n.* 容量
medium [ˈmiːdiəm]	*adj.* 中等的
plus [plʌs]	*prep.* 加
preset [ˌpriːˈset]	*vt.* 预置；预置的

extremely [ikˈstri:mli]		adv. 极端地；非常
complicated [ˈkɔmplikeitid]		adj. 结构复杂的
geometry [dʒiˈɔmətri]		n. 几何形状；几何图形
re-position [ˌri:pəˈziʃən]		vt. 重新定位

Phrases and Expressions

tool magazine	刀库
automatic pallet changer (APC)	自动工作台交换装置 (APC)
rotary table	旋转台
in addition to	除……之外
in conjunction with	联动；与……配合

Exercises

I. Read and judge.

(　　) 1. CNC machining centers developed from milling machines.

(　　) 2. All CNC machining centers have a tool magazine.

(　　) 3. The capacity of a magazine can be as large as several hundred tools.

(　　) 4. A four-axis vertical machining center has X-, Y-, Z- and B-axis.

(　　) 5. The most advanced CNC machining centers are designed to add two more axes (A-, B- or C-axis) in addition to the three basic axes (X-, Y- and Z-axis).

II. Translate the following phrases.

English	Chinese
1.	刀库
2.	刀具夹
3.	旋转台
4. automatic pallet changer (APC)	
5. a four-axis machining center	
6. in conjunction with	

III. Fill in the blanks according to the text.

1. A machining center is a mill with an _____ (ATC) that includes a _____, and sometimes an _____ (APC).

2. A four-axis machining center has X-, Y- and Z-axis as _____, plus a _____.

3. A machining center with four-axis control can automatically select and change as many as 32 _____.

4. The most advanced CNC milling centers are designed to add two more axes (*A*-, *B*- or *C*-axis) _____ the three basic axes (*X*-,*Y*- and *Z*-axis).

IV. Translate the following sentences into Chinese.

1. The medium size machining center may have 20~40 tools.

_____

2. Reduce machining time and cost.

_____

3. Better surface finish.

_____

4. No need to re-position the workpiece at complicated angles.

_____

5. Machine much more complicated parts, including normal holes required on a complicated surface.

_____

Unit 9　CNC Operation and Programming

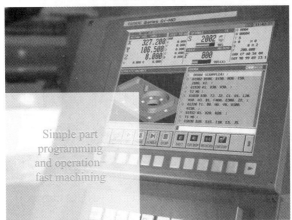

Simple part programming and operation— fast machining

In this unit, you will learn

◇ 1. Operation panel—system control panel;
◇ 2. Operation panel—machine control panel;
◇ 3. Commonly used codes and control functions.

Lesson 1 Operation panel—system control panel

An operation panel usually consists of two sections: the CNC system control panel and the machine control panel. The CNC system control panel also consists of two parts: the LCD (liquid crystal display) screen and soft keys and MDI (manual data input) keyboard. Here takes an operation panel with FANUC series 0i -MD (Fig. 9.1.1) for example.

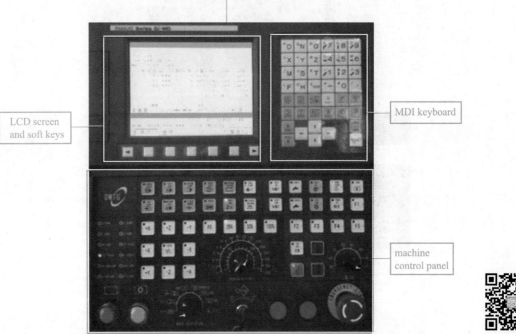

Fig.9.1.1 CNC operation panel with FANUC control

1. LCD screen and soft keys

The LCD screen shows the program, axis locations, and other information throughout the machine process.

The soft keys are below the screen and used to display more detailed functions.

2. MDI keyboard

The MDI keyboard (Fig.9.1.2) is the operation panel for the CNC control. It provides

Lesson 1 Operation panel—system control panel

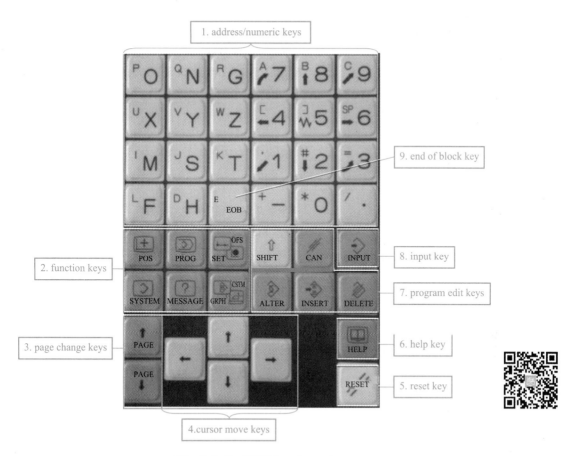

Fig.9.1.2 MDI keyboard

operating software for a CNC machine. It includes address/numeric keys, function keys, page change keys, reset key, etc. It allows the operators to enter commands by address/ numeric keys, and select functions needed by function keys.

2.1 Address/numeric keys

Press these keys to input letters, numeric, and other characters. For example, $\text{\tiny EOB}^{E}$ is end of block key, press this key and ";"will appear on the screen, which means the end of the block.

2.2 Function keys

There are six function keys: position key, program key, offset/setting key, system key, message key and custom/graph key, as described in the following table.

keys	Name	Function
POS	position key	Display the current coordinate position on the screen
PROG	program key	Display the program on the screen
OFS/SET	offset/setting key	This is a dual function key. Display offset/settings screen
SYSTEM	system key	Display system screen
MESSAGE	message key	Display the messages about the alarms and operations. It may also show alarm history
CSTM/CRPH	custom/graph key	Show tool path motions graphically

2.3 Program edit keys

There are five program edit keys: shift key, cancel key, alter key, insert key and delete key.

Keys	SHIFT	CAN	ALTER	INSERT	DELETE
Name	Shift key	cancel key	alter key	insert key	delete key

2.4 Some other keys and functions

Keys	Name	Function
PAGE↓ PAGE↑	page change keys	These two kinds of keys are used to change the page on the LCD screen forwards (↓) or backwards (↑)

continued

Keys	Name	Function
↑←↓→	cursor move keys	Move the cursor in different directions
INPUT	input key	Input the value
HELP	help key	Press the key to display how to operate the machine, or the details of an alarm
RESET	reset key	Reset the CNC system or cancel an alarm

New Words

9-1 单词

crystal [ˈkrɪstl]   n. 晶体；水晶
keyboard [ˈkiːbɔːd]   n. 键盘
reset [ˌriːˈset]   vt. 重置；复位
graph [græf]   n. 图表
alarm [əˈlɑːm]   n. 警报
shift [ʃɪft]   vt. & vi. 改变
delete [dɪˈliːt]   vt. 删除
offset [ˈɔfset]   vt. 补偿；偏置
setting [ˈsetɪŋ]   n. 装置；设置

Phrases and Expressions

liquid crystal display (LCD)   液晶显示屏
manual data input (MDI)   手动数据输入

Exercises

I. Translate the following phrases.

English	Chinese
1.	液晶显示屏
2.	程序键
3.	换页键
4.	功能键
5.	信息键
6. manual data input	
7. cursor move keys	
8. offset/setting key	
9. system key	
10. reset key	

II. Fill in the blanks according to the text.

1. An operation panel usually consists of two sections: _____ and _____.

2. The CNC system control panel also consists of two parts: _____ and _____.

3. There are six function keys: _____, _____, _____, _____, _____ and _____.

4. There are five program edit keys: _____, _____, _____, _____ and _____.

Lesson 1 Operation panel—system control panel

III. Fill in the blanks with name of the keys of the MDI keyboard.

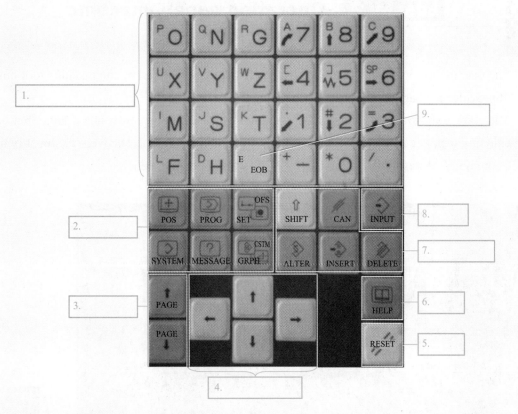

IV. Translate the following sentences into Chinese.

1. The LCD screen shows the program, axis locations, and other information throughout the machine process.

2. The soft keys are used to display more detailed functions.

3. The position key is used to display the current coordinate position on the screen.

4. Press end of block key and ";" will appear on the screen, which means the end of the block.

5. The reset key is used to reset the CNC system or cancel an alarm.

Lesson 2 Operation panel—machine control panel

The machine control panel (MCP) is designed and manufactured by the machine tool builder. It provides the hardware aspects of operating a CNC machine.

There are some small differences for operation of a machining center and a lathe, but both operation panels are similar. The following machine panel (Fig. 9.2.1) covers the most typical and common features found on modern FANUC series operation panel.

Fig.9.2.1 Machine control panel

Functions of the keys on a machine control panel are shown as follows.

No.	Keys/Switch	Name	Function
1		emergency stop (E-stop)	Press the button, and the machine is stopped emergently
2		(a) power on (b) power off	(a) The main power is turned on; (b) The main power is turned off

Lesson 2 Operation panel—machine control panel

continued

No.	Keys/Switch	Name	Function
3		mode selection	Select a mode, according to the operation: AUTO; EDIT; MDI; HANDLE; JOG; DNC—distributed numerical control; INC—incremental mode; REF—reference point mode
4	(a) (b)	(a) cycle start (b) feed hold	These are push buttons with lights. (a) Select an execution program and then press the button, the automatic operation is started. (b) Press the button during an automation operation, the tool speed decreases and then stops
5		indicator lights	Indicator lights indicate current control status of the machine. X HOME; SP. LOW: spindle low; SP. HIGH; spindle high; ATC READY: automatic tool changer ready; O.TRAVEL: overtravel; SP. UNCLAMP; AIR LOW; OIL LOW
6		jog axis selection	The keys are used to move the coordinate axes

continued

No.	Keys/Switch	Name	Function
7	(feedrate override dial)	feedrate override	Change the feedrate between 0~150%
8	(a) SPD ORI, (b) SPD CW, (c) SPD STOP, (d) SPD CCW	spindle rotation	(a) Spindle rotates original. (b) Spindle rotates clockwise. (c) Spindle stops. (d) Spindle rotates counter-clockwise
	(spindle speed override dial)	spindle speed override	Select the spindle rotary speed between 50%~120%
9	(a) CHIP CW, (b) CHIP CCW	chip cw chip ccw	(a) Chip device rotates clockwise to clean chips. (b) Chip device rotates counter-clockwise to clean chips
	(a) CLANT A, (b) CLANT B	clant (coolant)	(a) Press the button to turn on/off coolant A (fluid). (b) Press the button to turn on/off coolant B (mist)
	(a) ATC CW, (b) ATC CCW	tool magazine rotation	(a) Press the button to start the magazine in the clockwise direction. (b) Press the button to start the magazine in the counter-clockwise direction

Lesson 2 Operation panel—machine control panel

continued

No.	Keys/Switch	Name	Function
9	DOOR	door	Press the button to open/close the door
	WORK LIGHT	worklight	Press the button to turn on the working area lamp for enough light
10	SINGLE BLOCK	single block	Press the button on to allow the operator to process one block at a time
	DRY RUN	dry run	Dry run means machining the part dry. The word "dry" simply means machining without coolant. Dry run is used in the program testing stage where there is no part mounted in the fixture

New Words

9-2 单词

mode [məud]	n. 模式；方式
handle [ˈhændl]	n. 手轮；手柄
jog [dʒɔg]	n. 点动；手动进给
incremental [ˌiŋkrəˈmentl]	adj. 增加的；增量的
reference [ˈrefrəns]	n. 参考
execution [ˌeksiˈkju:ʃn]	n. 执行；实行
decrease [diˈkri:s]	n. 下降；减少
feedrate [ˈfi:dreit]	n. 进给速率；进给率
override [ˌəuvəˈraid]	n. 倍率
original [əˈridʒənl]	adj. 原始的
clockwise [ˈklɔkwaiz]	adv. 顺时针方向地
counter-clockwise [ˈkauntə klˈɔkwaiz]	adv. 逆时针方向地

fluid ['flu:id]   n. 液体
mist [mist]   n. 薄雾；液体喷雾
fixture ['fikstʃə]   n. 夹具；固定装置

Phrases and Expressions

emergency stop (E-stop)   紧急停止
incremental mode   增量模式
reference point mode   返回参考点
cycle start   循环启动
feed hold   进给保持
over travel (O.TRAVEL)   超程
single block   单程序段
dry run   空运行

Exercises

I. Give the full names of the following abbreviations and then translate them into Chinese.

1. E-stop　　　　＿＿＿＿＿＿＿＿＿＿＿＿＿　＿＿＿＿＿＿＿＿＿＿＿＿＿
2. MDI　　　　　＿＿＿＿＿＿＿＿＿＿＿＿＿　＿＿＿＿＿＿＿＿＿＿＿＿＿
3. INC　　　　　＿＿＿＿＿＿＿＿＿＿＿＿＿　＿＿＿＿＿＿＿＿＿＿＿＿＿
4. REF　　　　　＿＿＿＿＿＿＿＿＿＿＿＿＿　＿＿＿＿＿＿＿＿＿＿＿＿＿
5. ATC READY　＿＿＿＿＿＿＿＿＿＿＿＿＿　＿＿＿＿＿＿＿＿＿＿＿＿＿
6. SPD CW　　　＿＿＿＿＿＿＿＿＿＿＿＿＿　＿＿＿＿＿＿＿＿＿＿＿＿＿
7. CHIP CCW　　＿＿＿＿＿＿＿＿＿＿＿＿＿　＿＿＿＿＿＿＿＿＿＿＿＿＿
8. O.TRAVEL　　＿＿＿＿＿＿＿＿＿＿＿＿＿　＿＿＿＿＿＿＿＿＿＿＿＿＿

II. Translate the following phrases.

English	Chinese
1.	紧急停止
2.	循环启动
3.	进给保持
4.	空运行
5. single block	
6. incremental mode	
7. reference point mode	

III. Write out the functions of the following keys.

1. mode selection _____.
2. feed hold _____.
3. indicator lights _____.
4. jog axis selection _____.
5. feedrate override _____.
6. spindle speed override _____.

IV. Write the names of the following keys/buttons and translate them into Chinese.

Keys	1.	2.	3.	4. SPD CCW
Name				
Chinese				
Keys	5. CHIP CW	6. CLANT A	7. ATC CCW	8. SINGLE BLOCK
Name				
Chinese				
Keys	9. DRY RUN			
Name				
Chinese				

V. Translate the following sentences into Chinese.

1. Emergency stop—press the button, and the machine is stopped emergently.

_____

2. Cycle start—select an execution program and then press the button, the automatic operation is started.

_____

3. Spindle speed override—select the spindle rotary speed between 50%~120%.

4. Single block—press the button on to allow the operator to process one block at a time.

5. Dry run is used in the program testing stage where there is no part mounted in the fixture.

Lesson 3 Commonly used codes and control functions

1. Preparatory function

The preparatory function is to preset or prepare the CNC control system to a certain state of operation or to a certain mode. The instructions of preparatory function are represented by two digits followed by G. For example, G00 is rapid positioning (Fig.9.3.1a). The command will move a tool to the end position with an absolute or an incremental command at a rapid rate. G01 is linear interpolation (Fig. 9.3.1b).

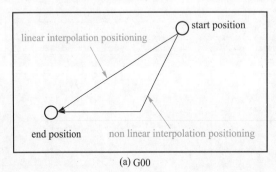
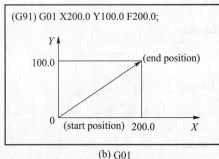

(a) G00 (b) G01

Fig.9.3.1 G00 and G01

The preparatory function indicates its meaning — a G-code will prepare control unit to accept the programming instructions following the G-code. For example, G00 presets a rapid motion mode for the machine tool but does not move. G81 (drilling cycle) presets the drilling cycle but does not drill any holes, etc.

For example, the block "N7 X13.0 Y10.0;" shows that coordinates relate to the cutting tool position, but does not indicate whether the coordinates are in an absolute or incremental mode. Neither does it indicate whether the actual motion to this position is a rapid motion or a linear motion. The information in such a block is incomplete.

In order to make the block N7 a tool motion in rapid mode (G00) using absolute dimensions (G90), all instructions must be "N7 G90 G00 X13.0 Y10.0;".

After the block number, G-codes are normally listed first in a block, followed by axes data, then all remaining instructions. For example, "N40 G91 G01 Z-0.625 F8.5;".

2. M function (Miscellaneous function)

The M function in a CNC program identifies a miscellaneous function.

Miscellaneous functions falls into two main groups: control of machine function and control of program execution.

The instructions of control of machine function control the spindle's rotation or stop, coolant's on and off, and tool's change, etc. For example, M03-spindle rotation normal (CW for right-hand tools), M04 -spindle rotation reverse (CCW for right-hand tools), M07 (fluid coolant on)(Fig.9.3.2a), M08 (mist coolant on) (Fig. 9.3.2b), etc.

The instructions of control of program execution control the end of program (M30), calling of subprogram (M98), etc.

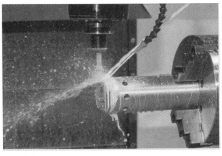
(a) M07

(b) M08

Fig.9.3.2   M07 and M08

The block "N25 G90 G00 X13.0 Y4.6 M08;" shows a rapid tool motion to the absolute position of (X13.0, Y4.6) within current settings and with a mist coolant turned on.

3. S function (Spindle speed function)

The S function controls the spindle speed. The spindle speed means spindle's revolutions per minute (r/min). For example, S1500 means the rotation speed of the spindle is 1 500 r/min.

Most machine spindles can be rotated in two directions—clockwise or counter-clockwise (Fig. 9.3.3). The S instruction should be given together with the spindle revolution instruction (M03 or M04).

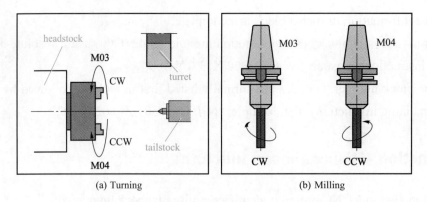

(a) Turning (b) Milling

Fig.9.3.3 Directions of spindle rotation

For example, the block "N230 S200 M03;" contains spindle speed of 200 r/min with spindle CW.

4. F function (Feed function)

The F function controls how fast the tool will move. The unit of tool feedrate is per minute or per revolution.

For the feed per minute (Fig. 9.3.4), the feed amount is the distance (millimeter) that a cutting tool will travel in one minute (60 seconds). F100 means the tool feedrate is 100 millimeters per minute.

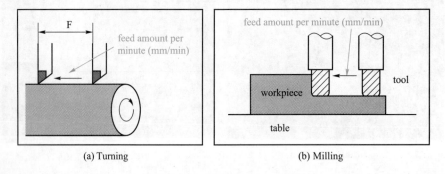

(a) Turning (b) Milling

Fig.9.3.4 Feed per minute

For the feed per revolution (Fig. 9.3.5), the feed amount is the distance (millimeter) that the tool travels in one spindle revolution.

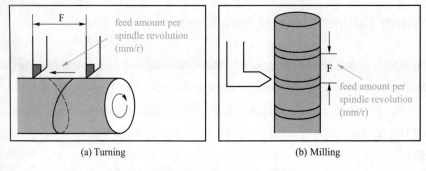

(a) Turning (b) Milling

Fig.9.3.5 Feed per revolution

There is a difference in G-codes used for lathes and machining centers.

feedrate	milling	turning
per minute	G94	G98
per revolution	G95	G99

For example, "G95 G01 X10 Y10 F0.1;" means the milling feedrate is 0.1 mm/r.

5. T function (Tool function)

For CNC machining centers, the T function (Fig. 9.3.6) controls the tool number only.

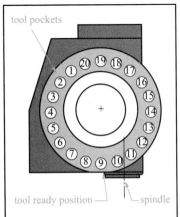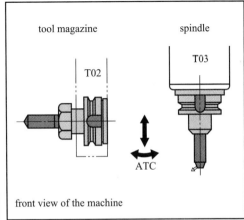

Fig.9.3.6 T Function of machining center

For example, T03 in "N67 T03" just indicates moving the tool to the waiting position.

Miscellaneous function M06 (tool change) must be used when actual tool change.

"N67 T03 M06;" means to bring tool No.3 into the spindle (tool change).

For CNC lathes, the T function controls indexing to the tool station number, as well as the tool offset number (Fig. 9.3.7). For example, T function T0101 will select tool station number one, and the associated tool wear offset of number one.

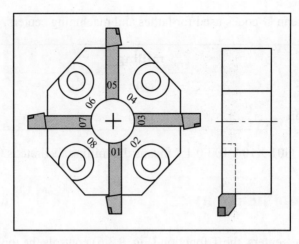

Fig.9.3.7　Typical lathe tool post (tool 1~8)

New Words

preparatory [pri'pærətri]	adj. 预备的；准备的
digit ['didʒit]	n. 数字；位数
absolute ['æbsəlu:t]	adj. 绝对的
interpolation [inˌtə:pə'leiʃn]	n. 插补；插值法
miscellaneous [ˌmisə'leiniəs]	adj. 辅助的；多方面的
subprogram ['sʌbprəugræm]	n. 子程序
revolution [ˌrevə'lu:ʃn]	n. 旋转
amount [ə'maunt]	n. 量；总额
millimeter ['miliˌmi:tə]	n. 毫米
index ['indeks]	vt. 索引；给……编索引
associated [ə'səuʃieitid]	adj. 联合的；有关联的

Phrases and Expressions

preparatory function	准备功能
rapid positioning	快速定位
linear interpolation	直线插补
miscellaneous function	辅助功能
spindle speed function	主轴转速功能
feed function	进给功能
tool function	刀具功能
tool offset number	刀具补偿值

Exercises

I. Translate the following phrases.

English	Chinese
1.	准备功能
2.	主轴转速功能
3.	进给功能
4.	刀具功能
5. rapid positioning	
6. linear interpolation	
7. miscellaneous function	
8. tool offset number	

II. Fill in the blanks according to the text.

1. The preparatory function is to _____ to a certain state of operation or to a certain mode.

2. G00 is rapid positioning. The command will move a tool to the end position with an _____ or an _____ command at a rapid rate.

3. M function in a CNC program identifies a _____.

4. S function controls _____.

5. F function controls _____.

6. For CNC machining centers, T function controls _____.

7. For CNC lathes, the T function controls indexing to _____, as well as _____.

III. Write down the functions of the following codes and then translate them into Chinese.

1. G00 _____ _____
2. G01 _____ _____
3. G81 _____ _____
4. M03 _____ _____
5. M08 _____ _____
6. M30 _____ _____
7. M98 _____ _____

IV. Explain the following blocks.

1. N7 G90 G00 X13.0 Y10.0;

2. N25 G90 G00 X13.0 Y4.6 M08;

3. N230 S200 M03;

V. Translate the following sentences into Chinese.

1. The preparatory function is to preset or prepare the CNC control system to a certain state of operation or to a certain mode.

2. After the block number, G-codes are normally listed first in a block, followed by axes data, then all remaining instructions.

3. The instructions of control of machine function control the spindle's rotation or stop, coolant's on and off, and tool's change, etc.

4. The spindle speed means spindle's revolutions per minute (r/min).

5. For CNC lathes, the T function controls indexing to the tool station number, as well as the tool offset number.

Unit 10 Job Application

In this unit, you will learn

◇ 1. Job advertisement;
◇ 2. Personal resume;
◇ 3. Job interview.

Lesson 1 Job advertisement

Position: Electromechanical engineer
Location: Shanghai, China
Job description:

- Analyze equipment failures, determine cause, and make necessary adjustments and repairs.
- Move, install and adjust fixtures, motors and other electrical / mechanical equipment.
- Perform preventative maintenance services.
- Read blueprints, schematics, diagrams or technical orders.

Job requirements:

- College degree or above in Mechanical or Integration of Mechanics and Electrics.
- Related experience in an industrial or manufacturing field.
- Knowledge of PLC (Programmable Logic Controller), Siemens Controls, Fanuc Controls and computer operated machinery.
- Good written and oral communication skills.

Benefits:

- Competitive base pay.
- Excellent benefit package including medical.
- Short term and long term disability insurance and life insurance.
- Paid vacation.
- Other allowance.

Company description:

ABC Inc. is a leading global manufacturer of electromechanical devices with annualized sales of $3.6 billion. ABC has about 15 000 employees working at more than 130 production bases and more than 100 sales and service centers in China and around the world. For more information, please visit the company website.

New Words

advertisement [əd'və:tismənt]	*n.*	广告
electromechanical [i'lektrəumi'kænikəl]	*adj.*	机电的；电动机械的
location [ləu'keiʃn]	*n.*	位置

Lesson 1 Job advertisement

preventative [pri'ventətiv]　　　　　*adj.* 预防性的
maintenance ['meintənəns]　　　　　*n.* 维修；维护
blueprint ['blu:print]　　　　　　　　*n.* 蓝图；设计图
schematics [ski'mætiks]　　　　　　*n.* 电路图
benefit ['benifit]　　　　　　　　　　*n.* 福利；津贴
insurance [in'ʃuərəns]　　　　　　　*n.* 保险
allowance [ə'lauəns]　　　　　　　　*n.* 津贴；补贴
annualized ['ænjuəlaizd]　　　　　　*adj.* 按年计算的
employee [im'plɔii:]　　　　　　　　*n.* 雇员

Phrases and Expressions

base pay　　　　　　　　　　　　　基本工资
benefit package　　　　　　　　　　福利待遇
disability insurance　　　　　　　　残疾收入保险
life insurance　　　　　　　　　　　人寿保险
paid vacation　　　　　　　　　　　带薪休假

Exercises

I. Match the words with their Chinese translations.

	A	B
( )	1. adjustment	a. 蓝图；设计图
( )	2. maintenance	b. 调试；调整
( )	3. insurance	c. 维修；维护
( )	4. blueprint	d. 雇员
( )	5. annualized	e. 保险
( )	6. diagram	f. 按年计算的
( )	7. employee	g. 示意图
( )	8. schematics	h. 固定装置
( )	9. fixture	i. 津贴；补贴
( )	10. allowance	j. 电路图

II. Translate the following words or phrases.

English	Chinese
1. preventative maintenance	
2. communication skill	

	continued
English	**Chinese**
3. disability insurance	
4. electromechanical engineer	
5.	福利待遇
6.	基本工资
7.	人寿保险
8.	带薪休假

III. Translate the following sentences into Chinese.

1. Perform preventative maintenance services.
   _____

2. College degree or above in Mechanical or Integration of Mechanics and Electrics.
   _____

3. Good written and oral communication skills.
   _____

4. Competitive base pay.
   _____

5. For more information, please visit the company website.
   _____

Lesson 2   Personal resume

A resume is a written compilation of your education, work experience, credentials and accomplishments used to apply for jobs. A resume is typically sent with a cover letter that provides additional information on your skills and experience to apply for jobs.

Name	Liu Chang	**Gender**	Male
Ethnic Group	Han	**Date of Birth**	05/1997
Education	College	**Height**	178 cm
Major	Mechatronics	**Political Status**	League Member
Graduated School	Jinan Vocational College		

continued

Job Objective	All positions related to mechatronics		
Mobil	**********	Postal Code	******
E-mail	*** ***** ****@***.com		
Address	302, Building 2, No. 365, Heping Road, Jinan, Shandong, China		
Education Background	Sept. 2015—July 2018　Jinan Vocational College Sept. 2012—July 2015　Jinan No. 2 Vocational School		
Courses	Advanced Mathematics, Machine Drawing, Auto CAD, Pro/E, Multisim, PLC, MasterCAM, Single-chip tutorial, Hydraulic technology, CNC milling machine, CNC lathe, etc		
Qualifications	CAD/CAM, Advanced Level CNC Programming and Operation, Advanced Level NCRE Grade 2		
Reward	National Scholarship (The third prize)		
Work Experience	03-06/2017　Internship as an Assistant Mechanical Engineer China FAW Group Corporation 07-08/2017　As a salesman of electromechanical products Jinan Machine Tool Co.Ltd		
Hobbies	Basketball, Playing guitar		
Self-Evaluation	I take responsibility in my study and work. I have rich experience in CNC lathe work, CNC Milling work practice, and etc. I like the profession of drafting and CNC programming		

New Words

10-2 单词

resume [ˈrezəmei]　　　　　　　　　n. 简历；摘要
compilation [ˌkɔmpiˈleiʃn]　　　　　n. 编辑；汇编
credentials [krəˈdenʃlz]　　　　　　n. 凭证；证件
accomplishment [əˈkʌmpliʃmənt]　　n. 成就；技能
gender [ˈdʒendə]　　　　　　　　　n. 性别
height [hait]　　　　　　　　　　　n. 身高
major [ˈmeidʒə]　　　　　　　　　 n. 专业
mathematics [ˌmæθəˈmætiks]　　　　n. 数学
objective [əbˈdʒektiv]　　　　　　　n. 目标

qualification [ˌkwɔlifiˈkeiʃn]   n. 资质
scholarship [ˈskɔləʃip]   n. 奖学金
responsibility [riˌspɔnsəˈbiləti]   n. 责任；责任心

Phrases and Expressions

national computer rank examination (NCRE)   全国计算机等级考试
national scholarship   国家奖学金
First Automobile Works (FAW)   中国一汽

Exercises

Create your own resume.

Name		Gender		
Ethnic Group		Date of Birth		
Education		Height		
Major		Political Status		
Graduation School				
Job Objective				
Mobil		Postal Code		
E-mail				
Address				
Education Background				
Courses				
Qualifications				
Reward				
Work Experience				
Hobbies				
Self-Evaluation				

Lesson 3   Job interview

Peter has got an interview notification from ABC company. Mr. Smith, the manager of the company is interviewing him.

P=Peter Wang   S=Mr. Smith

S: Come in, please.

P: Good morning, Mr. Smith.

S: Good morning. Have a seat, please.

P: Thank you.

S: Are you Mr. Wang?

P: Yes, I am Peter Wang.

S: I have read your resume. What are you primarily interested in about mechanical engineering?

P: I like designing products, and one of my designs won an award. Moreover, I am familiar with CAD. I can do any mechanic related work well if I am employed.

S: What are your goals for the next couple of years?

P: I want to learn more about my major and my work, and to increase my responsibilities and skills. My goal is to be the best.

S: We have several applicants for this position. Why do you think you are the person who could be chosen?

P: Hard working is the key to success. I believe that working hard makes the difference. And I will make efforts to make sure that I can do the job better.

S: That sounds very good. Do you have any questions about the job?

P: Yes. Do you offer any opportunities for further study?

S: Yes. You are encouraged to take some training. If you finish the training successfully, all the expenses will be reimbursed by the company.

P: That's fine.

S: Well, thank you very much, Mr. Wang. I'll let you know the result of the interview as soon as possible. Goodbye.

P: Thank you, Mr. Smith. I do hope to join you. Goodbye.

10-3 单词

New Words

interview [ˈintəvjuː]	n. 面试
notification [ˌnəutifiˈkeiʃn]	n. 通知
primarily [praiˈmerəli]	adv. 主要地
employ [imˈplɔi]	vt. 雇用；使用
goal [gəul]	n. 目标；目的
applicant [ˈæplikənt]	n. 申请人；求职人
effort [ˈefət]	n. 努力
opportunity [ˌɔpəˈtjuːnəti]	n. 机会
expense [ikˈspens]	n. 费用；花费的钱；消耗
reimburse [ˌriːimˈbəːs]	vt. 偿还；补偿

Phrases and Expressions

be familiar with ...	对……熟悉
a couple of	几个
the key to the success	成功的关键
make the difference	起（重要）作用，有影响
make sure	设法确保

Exercises

I. Read and judge.

(   ) 1. Peter is familiar with CAD.

(   ) 2. Peter likes drawing machines.

(   ) 3. Working hard makes the difference.

(   ) 4. There are no opportunities for further study in ABC company.

(   ) 5. If you successfully finish the training, all the expenses will be reimbursed by the company.

II. Translate the following words or phrases.

English	Chinese
1. interview notification	
2. the key to success	

English	Chinese
3. further study	
4. opportunity	
5.	设法确保
6.	起重要作用
7.	申请人
8.	责任感

III. Fill in the blanks according to the text.

1. Peter has got an _____ from ABC company.

2. What are you primarily interested in about _____?

3. Do you offer any opportunities for _____?

4. I'll let you know the _____ as soon as possible.

IV. Translate the following into Chinese.

1. I like designing products, and one of my designs won an award.
_____

2. My goal is to be the best.
_____

3. Hard working is the key to success.
_____

4. I'll let you know the result of the interview as soon as possible.
_____

Word List

A		application	U1L1
abbreviation	U4L3	arc	U2L3
absolute	U9L3	articulated	U6L2
access	U2L3	assembly	U1L1
accessory	U6L1	associated	U9L3
accomplishment	U10L2	automated	U6L2
accumulator	U4L3	automatically	U1L1
actuator	U6L3	automotive	U1L3
adjustment	U8L2	auxiliary	U8L2
advertisement	U10L1	award	U7L1
alarm	U9L1	axial	U1L2
algorithm	U5L2	axis	U7L1
allowance	U10L1	B	
alloy	U1L3	balance	U1L3
alter	U5L3	battery	U3L1
alteration	U5L3	bearing	U1L2
alternate	U3L2	benefit	U10L1
aluminum	U1L3	beverage	U6L3
ammeter	U3L2	blow	U5L3
amount	U9L3	blueprint	U10L1
ampere	U3L2	bolt	U1L2
amplifier	U3L1	boredom	U6L1
analogue	U4L1	boring	U7L2
analytical	U4L2	brass	U1L3
analyze	U3L3	bulb	U3L2
annualized	U10L1	bushing	U2L2
appliance	U3L2	byte	U4L1
applicant	U10L3		

continued

C		complicated	U8L3
cabinet	U5L1	component	U1L1
cable	U1L3	computerized	U7L1
can	U6L3	conductor	U3L1
capacitance	U3L1	constant	U4L3
capacitor	U3L1	constantly	U1L1
capacity	U8L3	construction	U2L2
carbon	U1L3	contract	U7L1
carriage	U8L1	control	U1L1
Cartesian	U6L2	controller	U1L1
cast	U1L3	convert	U5L1
cavity	U6L2	converter	U3L3
chain	U1L3	cookware	U1L3
characteristic	U6L3	coolant	U7L3
charge	U3L1	coordinate	U6L2
chassis	U5L3	copper	U1L3
chip	U4L1	core	U6L3
chromium	U1L3	corrosion	U1L3
chuck	U8L1	counter-clockwise	U9L2
circle	U2L3	coupling	U1L2
circuit	U1L3	create	U2L3
clamp	U1L2	credentials	U10L2
clarification	U2L1	crystal	U9L1
clockwise	U9L2	cube	U2L1
coating	U6L3	current	U3L1
code	U2L2	cursor	U2L3
cognition	U6L1	cushion	U1L2
coil	U7L2	cylindrical	U1L2
command	U2L3	D	
commercial	U5L1	damage	U7L3
tion	U10L2	datapath	U4L1
	U3L2	debug	U3L3
	U6L2	decrease	U9L2

continued

decrement	U4L3	element	U1L2
de-energize	U5L3	emergency	U6L1
defective	U5L3	emit	U3L1
delay	U5L3	employ	U10L3
delete	U9L1	employee	U10L1
density	U1L3	endurance	U6L3
description	U2L2	energy	U1L2
detail	U2L2	engineering	U1L1
diagnostics	U5L2	engrave	U7L2
diagram	U5L2	enlargement	U2L1
dielectric	U3L1	entire	U2L2
digit	U9L3	environment	U6L1
dimension	U2L2	erasable	U4L1
dimensional	U1L1	evaluate	U3L3
diode	U3L1	execute	U5L2
directly	U3L2	execution	U9L2
discharge	U3L1	expense	U10L3
display	U2L3	explosion	U6L3
distribution	U5L3	external	U2L2
divide	U1L3	extremely	U8L3
drafting	U2L3	**F**	
drill	U1L3	fabricate	U7L2
drilling	U7L2	facing	U8L1
ductile	U1L3	Farad	U3L1
E		fastener	U1L2
effector	U6L2	fault	U5L2
effort	U10L3	feed	U8L2
elastic	U1L2	feedrate	U9L2
electrical	U1L1	ferrous	U1L3
electrode	U7L2	file	U2L3
electromechanical	U10L1	fixture	U9L2
electron	U3L2	flame	U7L2
electronics	U1L1	flexibility	U8L2

continued

flexible	U1L2	implement	U5L1
flow	U1L1	include	U1L1
fluid	U9L2	increment	U4L3
formability	U1L3	incremental	U9L2
formula	U3L2	index	U9L3
frequency	U4L2	indication	U5L3
fuel	U7L3	indicator	U5L3
fused	U5L3	individual	U2L2
G		inductor	U3L2
gear	U1L2	inspection	U2L2
gender	U10L2	instruction	U2L2
generator	U3L2	instrument	U3L3
geometry	U8L3	insulate	U3L2
goal	U10L3	insulator	U3L1
graph	U9L1	insurance	U10L1
graphical	U5L2	integration	U1L1
grinder	U7L2	intelligence	U5L2
grooving	U8L1	intelligent	U1L1
H		intensity	U3L2
handle	U9L2	interconnection	U5L2
harsh	U6L3	interface	U2L3
headstock	U8L1	interfere	U1L2
height	U10L2	internal	U1L2
helical	U8L1	interpolation	U9L3
hip	U6L2	interpret	U5L1
hobbyist	U4L1	intersection	U1L1
horizontal	U8L2	interview	U10L3
housekeeping	U5L2	inversely	U3L2
humanoid	U6L2	invisible	U2L1
hydraulic	U7L3	involve	U6L2
I		iron	U1L3
icon	U2L3	item	U2L1
identify	U5L3		

continued

J		mechatronics	U1L1
jog	U9L2	medium	U8L3
joint	U6L2	menu	U2L3
junction	U7L3	metal	U1L3
K		microcontroller	U4L1
keyboard	U9L1	micrometer	U8L2
L		mill	U7L2
laboratory	U3L3	millimeter	U9L3
ladder	U5L2	military	U7L1
lathe	U7L2	miniature	U6L1
lettering	U7L2	miniaturization	U4L2
linear	U6L2	minimize	U3L3
liquid	U7L2	miscellaneous	U9L3
load	U3L2	mist	U9L2
locate	U6L3	mnemonic	U4L3
location	U10L1	mode	U9L2
logic	U3L3	model	U1L1
lubricant	U7L3	module	U4L1
M		moist	U7L3
magazine	U8L3	monitor	U1L1
magnetic	U1L3	mount	U1L2
magnify	U3L1	multifunctional	U1L1
mains	U5L1	multimeter	U3L3
maintain	U1L1	multiple	U6L1
maintenance	U10L1	multiply	U4L3
major	U10L2	**N**	
manipulator	U6L3	nail	U1L3
manufacture	U1L1	negative	U3L1
material	U1L3	non-ferrous	U1L3
mathematics	U10L2	non-metal	U1L3
maximize	U3L3	normal	U5L3
mechanical	U1L1	notification	U10L3
mechanics	U1L1	numerical	U7L1

continued

nut	U1L2	polyline	U2L3
O		port	U4L1
objective	U10L2	positive	U3L1
obstacle	U6L2	potential	U3L2
offset	U9L1	precision	U6L1
Ohm	U3L2	predict	U6L1
opcode	U5L2	preparatory	U9L3
operand	U4L3	preset	U8L3
operation	U1L1	pressure	U1L1
opportunity	U10L3	prevent	U7L3
orbit	U6L3	preventative	U10L1
orientation	U8L2	primarily	U10L3
original	U9L2	prismatic	U6L2
originate	U3L2	process	U1L1
orthographic	U2L1	processor	U4L1
override	U9L2	project	U2L1
P		projection	U2L1
package	U6L3	prompt	U2L3
pad	U3L1	proof	U6L3
pallet	U8L3	property	U1L3
palletize	U6L3	proportional	U3L2
panel	U2L3	pulley	U1L2
parallel	U8L2	punch	U7L2
parameter	U7L3	**Q**	
partial	U2L1	qualification	U10L2
parting	U8L1	**R**	
perform	U1L1	radiation	U6L1
peripheral	U4L1	raw	U7L2
perpendicular	U6L2	rear	U2L1
physicist	U3L2	recover	U7L3
plane	U2L1	reference	U9L2
plus	U8L3	refrigerator	U4L2
pocket	U8L3	register	U4L3

continued

regulation	U2L1	selective	U6L2
reimburse	U10L3	semiconductor	U3L1
relationship	U3L2	sequence	U5L1
reprogrammable	U1L1	sequential	U5L2
relay	U5L1	servomechanism	U7L1
remove	U8L1	setting	U9L1
repeatability	U7L1	shaft	U1L2
repeatable	U6L1	shaper	U8L2
replacer	U5L1	shear	U7L2
re-position	U8L3	shift	U9L1
resemble	U5L2	shoulder	U6L2
reset	U9L1	signature	U2L2
resistance	U1L3	simplify	U1L1
resistor	U3L1	simulate	U3L3
resource	U2L3	simulation	U3L3
responsibility	U10L2	sleeve	U2L2
responsible	U2L2	slot	U8L2
resume	U10L2	software	U1L1
revolution	U9L3	solder	U3L1
ribbon	U2L3	source	U3L2
rigid	U1L2	spark	U7L2
robotics	U1L1	specification	U2L2
rotate	U1L2	spherical	U6L2
S		spindle	U8L1
scale	U2L1	spreadsheet	U4L1
scan	U5L2	spring	U1L2
schematic	U3L3	stack	U4L3
schematics	U10L1	stainless	U1L3
scholarship	U10L2	standard	U2L1
screw	U1L2	status	U2L3
screwdriver	U1L2	steel	U1L3
sealant	U6L2	submerge	U7L2
section	U2L1	subprogram	U9L3

continued

subtract	U4L3	turning	U8L1
superintelligent	U6L1	**U**	
support	U1L2	unauthorized	U5L3
surface	U8L1	undo	U2L3
switch	U3L1	universal	U6L1
switchboard	U4L2	unsuitable	U6L2
symbol	U2L2	**V**	
system	U1L1	valve	U1L3
T		variable	U1L1
tab	U2L3	versatile	U6L2
tailstock	U8L1	vertical	U6L1
tamper	U5L3	view	U2L1
temperature	U4L2	virtual	U1L1
temporary	U5L2	visible	U2L1
terminal	U3L1	volt	U3L2
thread	U1L2	voltage	U3L1
threading	U8L1	voltmeter	U3L2
toggle	U4L1	volume	U7L1
tool	U1L1	**W**	
toolbar	U2L3	walkie-talkie	U4L2
transfer	U2L3	washer	U1L2
transformer	U7L3	weldability	U1L3
transistor	U3L1	wrench	U1L2
transmit	U1L2	**Z**	
troubleshooting	U5L3	zinc	U1L3

Phrases and Expressuions

A	
a couple of	U10L3
air conditioner	U4L2
alternating current (AC)	U3L2
analogue to digital converter module	U4L1
anti-lock braking system (ABS)	U4L2
application menu	U2L3
arc or circular movement	U8L2
arithmetic instruction	U4L3
assembly drawing	U2L2
automatic pallet changer (APC)	U8L3
automatic tool changer (ATC)	U8L2
auxiliary linear movement	U8L2
B	
ball bearing	U1L2
base pay	U10L1
be composed of	U1L1
be familiar with sth	U10L3
be similar to	U8L1
benefit package	U10L1
bit-oriented instruction	U4L3
boring machine	U7L2
branch instruction	U4L3
building automated communications call system	U4L2
C	
C language programming	U5L2
Cartesian coordinate system	U6L2
Cartesian robot	U6L2
central processing unit (CPU)	U4L1
coil winder	U7L2

continued

command prompt area	U2L3
component toolbar	U3L3
computer numerical control (CNC)	U7L1
computer aided design (CAD)	U1L1
computer aided testing (CAT)	U1L1
computer integrated manufacturing (CIM)	U1L1
consist of	U1L2
corrosion resistance	U1L3
cross cursor	U2L3
cutting off	U8L1
cycle start	U9L2
cylindrical grinder	U7L2
D	
data acquisition system	U4L2
data transfer instruction	U4L3
design toolbar	U3L3
detail drawing	U2L2
die-casting machine	U6L2
direct current (DC)	U3L2
directly proportional	U3L2
disability insurance	U10L1
drilling machine	U7L2
drive system	U8L1
dry run	U9L2
E	
electric charge	U3L2
electrically erasable programmable read only memory (EEPROM)	U4L1
electromagnetic interference (EMI)	U5L3
emergency stop button	U7L3
engraving systems	U7L2
explosion-proof performance	U6L3
F	
factory assembly line	U4L2

continued

first automobile works (FAW)	U10L2
feed function	U9L3
feed hold	U9L2
finishing pass	U8L1
flame cutting machine	U7L2
flat surface	U8L2
flexible automatic assembly system	U6L3
flexible coupling	U1L2
formed surface	U8L2
function block diagram	U5L2
G	
gear-cutting machine	U8L2
General Motors (GM)	U5L1
global position system (GPS)	U4L2
GPS navigation system	U4L2
H	
helical groove	U8L1
horizontal milling machine	U8L2
household appliance	U4L2
hydraulic pressure	U6L3
I	
in addition to	U8L3
in conjunction with	U8L3
incremental mode	U9L2
infocenter	U2L3
instruction list	U5L2
instrument toolbar	U3L3
insulating liquid	U7L2
intelligent control	U1L1
inversely proportional	U3L2
J	
jet nipple	U7L3
Joule heat	U3L2

continued

junction box	U7L3
L	
ladder diagram	U5L2
ladder logic	U5L2
Landis	U5L1
life insurance	U10L1
light-emitting diode (LED)	U3L1
linear interpolation	U9L3
linear movement	U8L2
liquid crystal display (LCD)	U9L1
load side	U5L3
logic converter	U3L3
logic instruction	U4L3
M	
machine control unit	U8L1
machine tool	U7L1
machining center	U7L2
main bar	U8L1
main view	U2L1
make sure	U10L3
make the difference	U10L3
manual data input (MDI)	U9L1
Massachusettes Institute of Technology (MIT)	U7L1
mechanical drawing	U2L1
menu bar	U3L3
microcontroller unit (MCU)	U3L3
milling cutter	U8L2
miscellaneous function	U9L3
Modicon Company	U5L1
mounting dimension	U2L2
N	
national scholarship	U10L2
national computer rank examination (NCRE)	U10L2

continued

O	
over travel (O. TRAVEL)	U9L2
orthographic projection	U2L1
oxide layer	U1L3

P	
paid vacation	U10L1
painting production line	U6L3
palletizing robot	U6L3
partial enlargement view	U2L1
PASCAL	U5L2
picking and placing	U6L2
power station	U3L2
preparatory function	U9L3
principal axis	U7L3
printed circuit board	U1L3
program-controlled switchboards	U4L2
programmable logic controller (PLC)	U5L1
projection view	U2L1
protective door	U7L3

Q	
quick access toolbar	U2L3

R	
random access memory (RAM)	U4L1
rapid positioning	U9L3
read only memory (ROM)	U4L1
reference point mode	U9L2
rigid coupling	U1L2
robot technology	U1L1
robot welding	U6L3
rotary table	U8L3
rough cut pass	U8L1

S	
safety operation	U7L3

continued

section view	U2L1
selective compliant assembly robot arm	U6L2
sequential function chart	U5L2
serial I/O module	U4L1
servo system	U7L1
short-circuit	U3L2
simulation run toolbar	U3L3
single block	U9L2
single-chip microcomputer (SCM)	U4L1
soldering machine	U7L2
specification dimension	U2L2
spherical robot	U6L2
spherical robot	U6L2
spindle speed function	U9L3
spot welding	U1L1
spray painting	U1L1
spray painting robot	U6L3
status bar	U2L3
structured text	U5L2
system toolbar	U3L3
T	
technical requirements	U2L2
terminal actuator	U6L3
the humanoid robot	U6L2
the key to the success	U10L3
three-view drawing	U2L1
title bar	U3L3
three- dimensional (3D) drawing	U1L1
timer module	U4L1
title block	U2L2
tool function	U9L3
tool holder	U8L2
tool magazine	U8L3

	continued
tool offset number	U9L3
tool post	U8L1
troubleshooting considerations table	U5L3
trunked mobile radio	U4L2
turning center	U7L2
two- dimensional (2D) drawing	U1L1
U	
United states Air Force	U7L1
V	
vertical milling machine	U8L2
video recorder	U4L2
W	
wire cut electrical discharge machine	U7L2
word processing	U4L1
workpiece holder	U8L1

References

[1] Cetinkunt S. Mechatronics [M]. New Jersey: John Wiley & Sons, Inc., 2015.

[2] Snyder D G. Introduction to Multisim for the DC/AC Course [M]. New York: Pearson Education, Inc., 2010.

[3] Shih H R. AutoCAD 2010 Tutorial [M]. Mission KS: Schroff Development Corporation, 2010.

[4] Fitzpatrick M. Machining and CNC Technology [M]. New York: The McGraw-Hill, 2013.

[5] Barrett F S. & Pack J D Microcontroller Programming and Interfacing [M]. Williston: Morgan & Claypool, 2011.

[6] 黄星,赵忠兴. 机电一体化专业英语[M]. 北京:人民邮电出版社,2013.

[7] 汤彩萍. 数控技术专业英语[M]. 北京:电子工业出版社,2013.

[8] 石金艳,谢永超. 机电与数控专业英语[M]. 北京:清华大学出版社,2014.

[9] 韩林烨,关雄飞. 机械类专业英语[M]. 北京:机械工业出版社,2013.

郑重声明

高等教育出版社依法对本书享有专有出版权。任何未经许可的复制、销售行为均违反《中华人民共和国著作权法》，其行为人将承担相应的民事责任和行政责任；构成犯罪的，将被依法追究刑事责任。为了维护市场秩序，保护读者的合法权益，避免读者误用盗版书造成不良后果，我社将配合行政执法部门和司法机关对违法犯罪的单位和个人进行严厉打击。社会各界人士如发现上述侵权行为，希望及时举报，本社将奖励举报有功人员。

反盗版举报电话　　（010）58581999　58582371　58582488
反盗版举报传真　　（010）82086060
反盗版举报邮箱　　dd@hep.com.cn
通信地址　　北京市西城区德外大街4号
　　　　　　高等教育出版社法律事务与版权管理部
邮政编码　　100120

防伪查询说明

用户购书后刮开封底防伪涂层，利用手机微信等软件扫描二维码，会跳转至防伪查询网页，获得所购图书详细信息。也可将防伪二维码下的20位密码按从左到右、从上到下的顺序发送短信至106695881280，免费查询所购图书真伪。

反盗版短信举报

编辑短信"JB,图书名称,出版社,购买地点"发送至10669588128

防伪客服电话

（010）58582300

学习卡账号使用说明

一、注册/登录

访问http://abook.hep.com.cn/sve，点击"注册"，在注册页面输入用户名、密码及常用的邮箱进行注册。已注册的用户直接输入用户名和密码登录即可进入"我的课程"页面。

二、课程绑定

点击"我的课程"页面右上方"绑定课程"，正确输入教材封底防伪标签上的20位密码，点击"确定"完成课程绑定。

三、访问课程

在"正在学习"列表中选择已绑定的课程，点击"进入课程"即可浏览或下载与本书配套的课程资源。刚绑定的课程请在"申请学习"列表中选择相应课程并点击"进入课程"。

如有账号问题，请发邮件至：4a_admin_zz@pub.hep.cn。